FALSE PROPHETS

FALSE PROPHETS

Alexander Kohn

Basil Blackwell

© Alexander Kohn 1986

First published 1986
Reprinted 1987

Basil Blackwell Ltd
108 Cowley Road, Oxford OX4 1JF, UK

Basil Blackwell Inc.
432 Park Avenue South, Suite 1503,
New York, NY 10016, USA

British Library Cataloguing in Publication Data

Kohn, Alexander
False prophets: fraud and error in
science and medicine.
1. Errors, Scientific   2. Impostors and
imposture
I. Title
500   Q112.5.E77
ISBN 0-631-14685-7

Library of Congress Cataloging in Publication Data

Kohn, Alexander.
False prophets.
Bibliography: p.
Includes index.
1. Fraud in science.   I. Title.
Q172.5.F7K64   1986   500   86-4254
ISBN 0-631-14685-7

Typeset by DMB (Typesetting), Oxford
Printed in Great Britain by
Billing and Sons Ltd, Worcester

Contents

Acknowledgements

The author and publishers acknowledge with gratitude permission to reproduce the following pictures: Babbage, Ptolemy, Newton and Mendel: Mary Evans Picture Library; midwife toad: from A. Koestler (1972) *The Case of the Midwife Toad*, New York, Random House; Fletcher: *Physics Today* (1982) June, 43, American Institute of Physics; Summerlin and his mouse: © J. P. Laffont, Sygma, from G. McBride (1974) *Journal of the American Medical Association* 229: 1391; Spector: *New Scientist* (1981); Felig: Owen Franken, Sygma; Piltdown skull: Department Library Services, American Museum of Natural History; Shapira figurines: Mrs Shulamit Lapid, Tel Aviv; Etruscan sarcophagus: Hebrew Encyclopedia, Jerusalem; Ferguson papers: *Science* (1981) 214: 1135 and *Experientia* (1981) 37: 252; Lock's remarks: *BMJ* (1984) 288: 662, 'Tricks of the Testing Trade': from M. Dowic et al. (1982) *Mother Jones*, Foundation for National Progress, June, 38–49.

Preface

Woe to the foolish prophets who follow their own spirit and have seen nothing.

(Ezekiel, 13: 3)

Breaches of ethics as encountered in scientific research cover a whole spectrum ranging from outright fraud and conscious falsification, through plagiarism and concealment of information, to minor infractions such as 'grantsmanship' and negligence.

There are differences of opinion concerning the seriousness of the problem posed by dishonesty in science. The traditional view among scientists, to which I subscribe, is that if falsehood is important enough to have damaging consequences it will be revealed soon enough, either via the unsuccessful attempts of others to replicate the work, or by inside information from the laboratory concerned. On the other hand, frauds involving experimentation without important consequences are by definition trivial and do little damage to science other than to clog up the scientific literature with the publication of fraudulent data.

This book deals with various aspects of misconduct in science – historical and recent – in the fields of the natural sciences, medicine, psychology and archaeology. It also includes accounts of erroneous research; these forms of aberration in science are often unintentional and are based on good faith. They may, nevertheless, contribute to the propagation of falsehood for quite some period of time, especially when initiated by scientists of good standing. I shall describe several examples of such erroneous research and explain their genesis and outcome.

The information has been gathered from primary scientific journals (Physical Review, Proceedings of the National Academy of Science,

Nature, Science, New England Journal of Medicine, Medical History etc.),
as well as from books and reviews. One chapter is devoted to drug
testing: the information for this was collected directly from the
archives of the Food and Drug Administration in Washington, DC.
The Freedom of Information Act has made it possible to screen all
the relevant documents mentioned in this chapter. The book is by no
means exhaustive, however; I have purposely omitted much confi-
dential information which has reached me in discussions with colleagues in various scientific disciplines.

The idea of writing this book arose in 1980, when I became interested in the modes of communal thinking of scientists while misled
by a non-existent 'discovery'. From there I was led to weigh the
effects of negligence on the outcome of scientific research, and finally
I began to consider how it happens that some scientists entirely relin-
quish the ethical rules of scientific conduct and engage in fraudulent
research. The final incentive came a few years ago when I encountered a PhD student who went to considerable trouble to falsify the
position of nucleic acid bands in an electrophoretic gel, though I am
not at liberty to describe this case.

I am indebted to many colleagues who discussed with me the various aspects of unethical behaviour in science, and provided me with
an additional insight into these problems as well as with source
material. I am further indebted to Mrs Janina Gitelman and Ms
Ruth Kohn for their technical help with the references.

1

Making and Breaking the
Rules of Science

The exponential growth of science and the increase in the number of practising scientists has been accompanied by the appearance of individuals whose actions do not conform with the ethical standards of the scientific community. Although the extent of this deviant behaviour has not yet been satisfactorily investigated and has still to be determined, there is a feeling of embarrassment and unease in the scientific community. This unease stems from the essential importance of honesty in science. Unlike other professions where honesty is merely regarded as highly desirable, the whole edifice of science is built upon honesty; it is not, as Jacob Bronowski has pointed out in *The Common Sense of Science*, an optional extra or a separate domain. 'The institution of science involves an implicit social contract between scientists so that each can depend on the trustworthiness of the rest . . . the entire cognitive system of science is rooted in the moral integrity of aggregates of individual scientists'.[1]

Scientists have therefore been deeply disturbed by the fact that during the past decade many incidents of plagiarism, data falsification, misrepresentation of research results and misuse of public funds for fraudulent research, have been reported in scientific periodicals as well as in the mass media. The latest contribution to the literature on this subject is a book by science reporters Broad and Wade,[2] entitled *Betrayers of the Truth*.

Before discussing the various forms of misconduct in science, however, we have to understand the basic rules, the norms of science.

Robert Merton, the pioneer of the sociology of science,[3] distinguishes between moral, intellectual and technical norms. He has classified these norms as 'universalism', 'disinterestedness', 'scepticism' and 'communalism'.

Universalism implies that truth should be judged in terms of intellectual criteria, criteria that are considered valid in the particular branch of science, and not in terms of the attributes of the author. Disinterestedness requires that the scientist's activities and efforts be directed towards the extension of scientific knowledge, and not towards the personal interests of an individual or a group of scientists. Scepticism involves testing claims both empirically and logically, and not accepting them on the basis of authority. Communalism requires that science be a product of collaborative effort, dictated by the wish to benefit the community, the society; it thus entails openness and sharing of information.

To these norms, as defined by Merton, one may also add rationality and emotional neutrality,[4] and according to Cournand and Zuckerman,[5] honesty, objectivity, tolerance, doubt of certitude and unselfish engagement. Mohr[6] presents the following list of normative rules: 'Be honest; never manipulate data; be precise; be fair with regard to priority; be without bias with regard to data and ideas of your rival; do not make compromises in trying to solve a problem.'

Kenneth S. Norris of the University of California, Santa Cruz, wraps it all up in a down-to-earth comment: 'Science is a set of rules that keep the scientists from lying to each other.'

Violation of these norms is sometimes loosely termed 'fraud'. Actually, in both everyday and legal language, fraud is defined as criminal deception, or as the use of false representation intended to benefit the deceiver. The scientist who is cheating knowingly, who falsifies or invents research data, or who lies about them, is not strictly fraudulent as long as he is not using the false data to obtain financial support from public or government agencies, or from private funds. Fraud is also committed when, on the basis of false data, the scientist is trying to secure a research job, to prove that public funds have been properly used, or to convince the public or the grantors that a certain procedure, material or drug is acceptable and safe. Fraud is more a problem of the social and biological sciences (the so-called 'soft' sciences) than the 'exact' sciences. This is because in the soft sciences one encounters uncontrollable variables which make the experiments difficult to reproduce; fraudulent results therefore have a better chance of cover-up.

Violation of the norms often gains public notice and is described in the lay press in an exaggerated manner. This creates the impression that such deviant behaviour is much more prevalent in the scientific community than is admitted by scientists themselves.[7] Considering the constantly increasing numbers of practising scientists, I would estimate that the number of deviant scientists is extremely small. Broad and Wade[2] consider fraud in science to be a small, but not insignificant feature of scientific research, and this view is supported by Dixon:[8] 'Blatant fiddling is probably an uncommon though real component of careerist science.' In serious and well-substantiated cases offending individuals, when found out, are punished by the scientific community by being cast out of the ranks.

During a period in the development of a particular science when clearly defined concepts and laws are widely accepted, it is difficult to cheat. When, however, a field of science enters a state of flux, with concepts changing, old patterns collapsing and new models trying to get hold, the area becomes most vulnerable to deceit. Objectivity becomes difficult, even to specialists. Science in revolution invites new hypotheses. At such transitional stages there are those who have the courage to promulgate unpopular hypotheses selflessly, but there are also opportunists who do not hesitate to flood the field with forged data and results of experiments that have not been carried out. In an atmosphere of change and flux, as we have witnessed in the past few decades, such incidents of forgery and misrepresentation are not isolated. During one such time of change when new concepts were emerging, when the new types of radiation were discovered at the turn of the century, 'discoveries' were also made of non-existent radiations (e.g. N-rays, mitogenetic rays); these were the result of negligent or badly controlled experimentation, however, not of deceit or conscious fraud.

Charles Babbage, the famous English mathematician (1792–1871) and the inventor of the first modern calculating machine, defined various types of scientific misconduct, or methods of leading people astray. One speaks of 'forging' when one records observations that have never been made; in other words, when one lies outright about the experimental data. When data are manipulated so as to make them look better, one deals with 'trimming' (in modern usage also 'massaging data' or 'fudging'). 'Cooking', as defined by Babbage, means choosing only data that fit the researcher's hypothesis best, and discarding those that do not: in this case one tells only a part of the truth (in modern jargon, 'finagling').

Let us examine examples of different types of cheating in contemporary science. 'Forging' involves reporting experiments that have never been carried out, but which the researcher may feel he needs for the support of his or her hypothesis. In some cases forgers have not only invented experimental data, but have related these non-existent data to non-existent manuscripts allegedly published or accepted for publication.

Another form of cheating in this category is plagiarism, which consists of using the data or ideas of other investigators without reference to the source, or even verbatim copying of a text written by someone else. A number of plagiarizers obtained the necessary data either from a grant proposal or from a manuscript submitted for publication.

In the 'trimming' category one would include that type of cheating which is based on amplification of an experiment: the investigator describes accurately the nature of the experiment and its controls, but reports a greater number of trials than have actually been performed, 'adds' or 'removes' animals to or from experimental or control groups, or misrepresents the variance, although using genuine numbers and means.

One encounters examples of 'cooking' when the researcher omits aberrant values, misreports actual conditions of the experiment or alters ancillary data. A kindred misdemeanour is omission of whole experiments which yielded negative or contrary results to the hypothesis under test.

A number of questions come to mind relating to the norms of science and deviant behaviour. Is there a system of social control in science and what is its nature? What is the incidence of deviant behaviour in science in comparison with that in other fields of human endeavour? Why is there no systematic effort to study such behaviour? Where should the line be drawn between an error and forgery or plagiarism? Is there any connection between the reward system in science and the incentive for deviant behaviour?[1] I shall hope to suggest answers to these questions in the course of this book.

Perhaps we should look first at the existence of 'social control' in science. One of the first lessons a scientist learns is that faking evidence is the worst sin he can commit; it is a sort of capital crime. He also learns that those who commit such crimes are banished from the scientific community, or at best, held in contempt, suffering resentment and antipathy. A deviant scientist will most likely be barred from holding a position in a research organization or university.

Scientists seek recognition: they want to publish and to see their names in print; they wish to be recognized by their fellow scientists, and if the recognition comes from the higher levels of the scientific hierarchy so much the better; they like to be invited to meetings and conferences, and the smaller the conference and the farther from home the better. Scientists fight fiercely for priority of their discoveries. Jevons[9] states that the major motives for a scientist are recognition from fellow scientists, a permanent drive to work, to be creative and not to violate the ethics of science, and, if possible, to be first with a new discovery.

Nevertheless, scientists may be tempted by power or fortune, even if they do not actually seek them, contrary to popular belief that most scientists are willing to work anonymously for the advancement of knowledge. The international scientific community evaluates and judges the work of its members – from publications, lectures, discussions and sometimes even by the grapevine. It controls the activities of individual scientists by granting recognition to the successful members, and withholding it from those who violate the norms of scientific behaviour.

Any new concepts, hypotheses or proofs which have a bearing on the mainstream of science are examined by other members of the scientific community. They are accepted, modified or rejected (sometimes even ignored). The highest reward for a scientist is peer recognition of the validity and importance of his or her particular truth-claim. Thus the organized system of scepticism has a double function: it improves the quality of scientific investigation and it reduces the extent of possible frauds. This scepticism assumes many forms. First there is the self-imposed careful statistical design of experiments which includes double blind tests, randomization of subjects, use of independent observers and, above all, the requirement that the experiment be reproducible.

The concept of the double blind test is extremely important in biological sciences. If a scientist assumes that procedure or substance A has a specific effect on micro-organisms, plants, or humans, he includes in his test procedure a substance B which is known to be 'neutral', without a demonstrable effect. The conditions of the test are such that A and B are coded and neither the experimenter (or his delegate) nor the subject knows whether A or B has been used in any particular test. The code is broken only after the result of the test has been read and recorded.

As to randomization, if a test is done within a population, the subjects of the study have to be chosen at random (regardless of sex, age, ethnic origin, behavioural traits such as smoking, drinking etc.). Selection of individuals for a study may lead the researcher to mistaken conclusions owing to unrecognized bias in the selection.[10]

The use of independent observers is important because each observer may interpret the criteria (in clinical studies, for example) differently and apply his individual interpretation to the collected information. One example of this in practice is the frequently varying or even contradictory opinions of pathologists evaluating pathological slides.

In spite of all these precautions, many a research project, especially in the field of psychology, is burdened by so-called 'experimenter bias' (see chapter 2). Even in well-designed and controlled experiments the outcome may be biased in favour of a preset hypothesis, set up by, or known to the experimenter. In addition, the number of repeat experiments a scientist performs is usually smaller when the results of an experiment conform to the hypothesis, than in a situation where the results are not as expected.

The attitude to reproducibility, a key factor in controlling fraud, differs in various sciences. It is usually more stringent in the exact sciences (such as physics and chemistry) than in the social and behavioural sciences. Problems arise because the fact that an experiment is not reproducible does not necessarily mean that the concept on which it is based is wrong, nor that there is an error or deception in it. Gravitational waves provide one example. Joseph Weber, an internationally known physicist from the University of Maryland, constructed special detector antennae to measure the existence of these waves, and positioned the antennae at a distance of a 1000 km from each other (in Maryland and Illinois). Through them he detected a coincident significant signal, which he interpreted as evidence for gravitational wave bursts coming from sources outside our solar system.[11] Though he regarded these signals as significant and excluded the possibility that they might be due to electromagnetic or seismic effects, other physicists considered them to be 'noise'. Later studies with much more sensitive apparatus have not produced any signals comparable to those observed by Weber. This fact, however, does not necessarily invalidate Weber's work and hypothesis, though it does make other workers suspend judgement until more experiments have been carried out.

Dirac once said that it was more important to have beauty in an equation than to have it conform to experiment. Nevertheless, whenever a number of laboratories encounter difficulties in reproducing an experiment, the original findings may fall under a shadow of doubt. In such cases efforts will be made to clarify the causes of irreproducibility. If all attempts fail, a suspicion of deception may arise and will usually be circulated in closed circles but not in public. Voicing of such suspicion is avoided by senior scientists who do not want to destroy the career of their junior colleagues; junior scientists do not dare to accuse their superiors or professors for more empirical reasons. In short, when faced with irreproducible results, we all tend rather to assume that an error has been made rather than a fraud committed. The attractiveness of any theory in its application to society depends on how relevant it is and how well it fits everybody's preconceptions. If the fit is good, then even if the proofs are shaky, there would be a tendency to accept the theory.

The next question we might ask ourselves is just how widespread is the kind of misconduct we have been talking about? At a meeting of the Society of Sociology of Science held in Philadelphia in 1982, Professor Robert Merton complained that there had been numerous discussions on misconduct in science, but that actual quantitative data on the incidence of fraud were still lacking. This situation remains unchanged and it is therefore very difficult both to state how prevalent deviant behaviour in science is, and to make comparisons with the incidence of fraud in other fields of human endeavour. In distinction from other crimes, deviant behaviour in science is not officially reported by the police or the law courts. Therefore there are no official statistics relating to this type of 'crime'. The main source of information is either self-reporting by individuals (admission of guilt in a scientific journal or at a meeting) or 'whistle-blowing', that is, disclosure of fraud by the scientist's colleagues, collaborators or technicians.

In 1976 the *New Scientist* embarked on a study of cheating in science.[12,13] This study was instigated by the affair of Sir Cyril Burt (chapter 4), who was at that time accused of having used non-existent data in his research on the inheritance of intelligence. *New Scientist* sent out questionnaires to its readers asking two main questions: (*a*) does intentional bias occur among scientists to the same extent as among non-scientists?; (*b*) how does one distinguish between deliberate misrepresentation and error or carelessness? The term 'intentional bias' was euphemistically used for intentional manipulation of data. The questionnaire contained additional questions, some of which dealt with the incidence of bias, others with knowledge of such practices.

Of 201 valid questionnaires returned to the editors, 194 reported having known about instances of cheating. About one-third of all cases had come to the knowledge of the reporters either from direct or indirect personal contact. Of all reported cases one-fifth of the perpetrators were actually caught in the act, and an additional fifth confessed to malpractice. The incidence of cheating was higher in research performed by single scientists and it lessened the larger the research group became. There were only 15 incidences of complete fabrication of an experiment that had never been carried out; the rest involved trimming and cooking. As to the fate of the 'intentional biasers' who were caught, only ten per cent were actually dismissed from their positions. Other forms of punishment are not described in this study.

It would thus seem that the prevalence of cheating in science is extremely high. It would obviously be a good idea, however, to conduct a much more extensive study of larger populations of practising scientists before accepting the conclusions of *New Scientist* in this matter.

An interesting study on the reliability of published scientific papers was carried out by Dr Richard R. Roberts of the National Bureau of Standards.[14] He estimated that at least half of all published scientific papers were unusable or unreliable, although this does not necessarily imply cheating.

Another type of test was initiated by Leroy Wolins of Iowa State University.[15] He authorized one of his students to write to 37 authors of psychological papers and ask them for the raw data on which they based their research results. Of the 32 who replied, 21 stated that their data had been either accidentally destroyed, lost, or misplaced. Only nine researchers sent their raw data. Dr Wolins, an expert in statistics, analysed these data and found that only seven sets of results could be statistically analysed. Of these seven, three contained errors that invalidated what had been published as fact.

The studies of *New Scientist*, of Roberts and of Wolins, taken together, would indicate that the prevalence of misconduct in science is greater than the scientific community is willing to admit, but these are very selected studies and do not cover the complete range of scientific activities. They should therefore not be used for extrapolation. The real prevalence of cheating may be smaller or greater.

In order to obtain a proper perspective on this problem, in the following chapters I shall analyse a number of published cases of scientific misdemeanour. First, however, consider the following hypothetical case. In an experiment, ten measurements are made of a parameter changing with time. The recorded values are plotted on a semilogarithmic paper as a straight line (i.e. exponential decay). The experiment is repeated five times. Within the limits of properly used statistical deviations, a straight line is indeed observed. In one of the five experiments, however, one of the points is so aberrant that the straight line relationship in this particular experiment cannot be maintained. The location of this point (rogue value) is statistically incompatible with the line drawn through all the other points.

We are now faced with a crucial question. The straight line relationship fits the particular hypothesis of the experimenter, who now wishes to write up the results and submit them for publication. For obvious reasons (limitation of space) he can use only one graph to demonstrate the point he is trying to make. In such a situation many scientists will choose the best results as an illustration. This would not be considered improper, since insistence on presenting the aberrant datum (the rogue value) would lead only to the obfuscation of the issue. Since most of the research done nowadays is cooperative, it is expected that results will have been discussed by all the partners, who would consider the pros and cons of stressing either the conforming values or the aberrant ones and come to a weighted conclusion based on experience, intuition and the communal mode of thinking. Nevertheless, in the strict ethical sense, the issue is debatable.

The course of action described above may still be considered by some purists as unethical. In practice, I would not condemn it as a misconduct. Few of these cases are 'open and shut'; they should be judged against a background of a great variety of factors.

Scientists who neglect the proscribed and sanctioned methodology (i.e., the use of proper controls) may quite unintentionally become responsible for producing erroneous results, and thus be suspected of falsification. A classic example of such methodologically faulty research was the discovery and the prolonged study of mitogenetic rays in the 1930s and of polywater in the 1960s (chapter 3). In the course of a decade several hundreds of papers on a non-existent ray and on a non-existent compound were published by reputable scientists, who failed to employ proper precautions in their research and in the interpretation of their results.

How can this have happened? Whenever a new scientific claim or discovery is made in a field that is important to the advance of knowledge, has practical technological applications or is of social importance, many people try to reproduce the work. As Merton put it, 'scientific inquiry is in effect subject to rigorous policing, to a degree unparalleled in any other field of human activity'.[3] Thus the requirement of reproducibility in scientific research aids in the detection of error or fraud, and is therefore a deterrent to misconduct. Nevertheless, there are situations in which repetition of an experiment is extremely difficult, because of the complicated experimental design, for instance, or because of its prohibitive cost, or because the phenomenon studied is rare and unexpected (such as earthquakes, volcano eruptions). In many cases, however, where the results of an experiment are not a part of a necessary link in the advance of knowledge and only serve to support a generally accepted 'truth', serious scientists will seldom be willing to lose time and money to repeat that particular experiment. Thus, many results that find their way into published literature remain unchecked, whether they are rarely or often quoted. In mathematics, for instance, according to one writer, 'this is a real problem: many published mathematical articles undoubtedly contain serious undetected errors, not because the mistakes are too difficult to find, but because contemporary pure mathematics has become so abstract and fragmented that few people bother to look carefully for error'.[16]

According to Luria,[17] actual cheating in science is the result of a pathological personality akin to that of a compulsive gambler; anyone in science who willingly distorts or invents data believing he will get away with it must have a disturbed sense of reality. To carry the comparison with a gambler further, such a scientist deludes himself that he can beat the odds and bend reality to his wishes. He has lost the awareness of the futility of cheating in science. A cheating scientist, like a gambler, is unable to respond to normal emotional impulses. 'Enthusiasm that brings the scientist to the laboratory door should be left outside along with umbrella and overshoes', says Luria, and should be replaced by scepticism. As another writer has put it: 'If science is misrepresented to the young as an array of glamorous "spectaculars" the seed is sown for later anticlimactic disillusionment and perhaps even dishonesty'.[18]

'Cheating may be like cirrhosis of the integrity. An initial slip made in good faith gives rise to a reaction that magnifies the emotional commitment to the mistaken belief, finally leading to the actual destruction of truth'.[17] Despite the centrality of honesty in the scientific enterprise, it is foolish to assume that dishonesty, common in most fields of human activity, should be rare in science. Falsification is inevitable in human society, be it scientific or otherwise. Science, like banking, however, has the ways and the means to keep misconduct in check.

2

Experimenter Effects

One of the major sources of error in research is the 'experimenter effect'. This concept is defined by Rosenthal[19] as 'the extent to which the datum obtained by an experimenter deviates from the correct "value". The measure of experimenter effects (or experimental error) is some function of the absolute (unsigned) deviations of that experimenter's data about the "correct value", i.e. a measure of total error.' One may add to this the opinion that all scientific inquiry is subject to error and that it is far better to be aware of it than to be ignorant of the errors concealed in the data.[20]

Consider an inexperienced market researcher who, on behalf of let us say the 'National Tourist Board', is polling the preferences of people to spend their holidays at home or overseas. Those polled respond inaccurately to the investigator's questions, depending on their personal characteristics and habits. The researcher may interpret the inaccurate opinions as answers favourable to his or her employers. Similar bias may be encountered in reports of research students desiring to please their supervisors while collecting evidence to support the theories or ideas of their professors.

THE PITFALLS OF BEHAVIOURAL RESEARCH

The experimenter effect is particularly prominent in behavioural research, where people exchange signals unintentionally, without speaking. These signals may be transmitted by gestures, by auditory or visual channels, by touch or even by smell. In every experimental situation the experimenter may thus convey to the subjects his (or her) feelings without even knowing that he has done so. Of the various types of experimenters defined by Rosenthal,[19] only the 'realistic' ones may suspect that their own research is affected by their expectations. Of the other two types, the 'incredulous' considers the concept of error as nonsense, and therefore does not pay any attention to the possibility that it might occur; the 'gleeful' knows all along that experiments in behavioural sciences are riddled with error.

How can the presence of the experimenter influence the investigated subjects? One technique that has been employed to elucidate this problem involves the use of photographs showing human faces expressing various emotions. In the test the subject is shown the photographs by the experimenter and asked to define whether the expression on the face indicates that the person had been experiencing failure or success. The subject then reports his evaluation to the experimenter present during the test. The conditions of the experiment (experience of the experimenter, the wording used by the experimenter while showing the pictures, the time schedules, the topography of the room where the tasks are given etc.) are varied from experiment to experiment. Rosenthal found that in spite of all the different conditions employed, the opinions expressed by the subjects continued to be influenced by the views of the experimenter.

One would like to think that numerous or frequent repetitions of an experiment would eliminate the observer errors, or at least diminish them. In fact, repetition of the same type of experiment does not eliminate the error since 'replicated observations made under similar conditions of anticipation, instrumentation and psychological climate may, by virtue of their intercorrelations, all be in error with respect to some external criterion'.[19]

Another erroneous belief is that the experimenter effect is applicable only to experiments with humans because unrecognized and unintentional signals may be exchanged only between humans. In fact, animals, too, are sensitive to gestures, tensions and other invisible or subconscious clues produced by their human masters or companions, as the case of the 'clever Hans' horse has amply demonstrated. This horse was supposed to know basic arithmetical operations: a person in the audience would pose a question such as 'What's 3×4?' and the horse would stamp its hoofs the correct number of times. Actually, the horse sensed a change in muscle tension of his trainer when the correct figure had been reached.

Rosenthal and his collaborators[21,22] carried out experiments designed to detect the experimenter effect on rats. In the first trial there were 12 experimenters: they put five white rats through a simple discrimination test daily for five days. In the following study there were 38 experimenters divided into 14 research teams. They had to train rats in seven different tasks. The experimenters were deliberately biased by having been provided with false information that some of the rats were 'bright' and the others were 'dull' while in fact all the rats were from the same colony, were of the same age and sex and had performed similarly. The 'intelligence' of the rats was said to have been determined in previous maze running experiments. Eight teams were given rats described as 'bright', and six teams were told their rats were 'dull'.

At the end of the experiment, the experimenters had to rate themselves, as well as the rats. It turned out that the experimenters believing their subjects to be generally 'bright' observed better performance on the part of the rats and rated themselves as more 'enthusiastic, friendly, encouraging, pleasant and interested' in connection with the performance of their rats, than the experimenters working with 'dull' rats. The differences between the two groups were statistically significant.

The explanation given to the experimenter effect in rats was that the rats defined as 'bright' and supposedly performing better, were liked better by their experimenters and were therefore touched more. Indeed, Bernstein showed in 1957[23] that rats learned better when they were handled more by the experimenters. If mere physical contact could affect the learning behaviour of rats surely more dramatic effects may be expected in human experimentation.

SEEING STARS

A significant number of cases attributable to experimenter bias have been reported in the field of astronomy. Re-examination of certain astronomers' published data indicated that the reported observations could not have been possible. John Flamsteed, an English astronomer of the seventeenth century, was the proponent of an (incorrect) theory predicting that the Pole Star would be closer to the North Pole in winter than in summer. He published his own astronomical observations to confirm his hypothesis; in other words, he found what he expected to find. When his colleagues later showed him that his observations had been erroneous, Flamsteed admitted the error and blamed it on the instruments, although he never succeeded in identifying the source of the error!

Another English astronomer, Thomas Harriot, in 1609 viewed the moon through a telescope. He drew the face of the moon as he observed it, but his drawings were closer to Galileo's map of the moon than to what Harriot could have seen with the telescopes then available.

A century later, William Herschel devoted his attention to the planet Uranus, which he assumed to be a comet approaching the solar system. He reported that the apparent size of the 'comet' increased during the period of his observations between March and April 1781. The truth, we now know, is that during that particular month the planet Uranus had been receding from Earth, and thus Herschel should have seen an apparent decrease and not, as he reported, an increase in size. In other words, what Herschel claimed to have seen had been impossible. I suggest his error may have been due to bias.

These three cases are all discussed by Hetherington,[24] who stated: 'Foreseen, anticipated or even desired results were found by scientists adhering to generally accepted scientific methods – even though we now know that the reported phenomena do not exist.'

Hetherington devoted special attention to several astronomers from the Mount Wilson Observatory near Pasadena in California who reported untenable data at the beginning of this century. Adrian van Maanen observed and photographed spiral nebulae. He detected in them internal motions by comparing photographs of the same nebula taken at different times.[25] From further observations, van Maanen concluded that the seven nebulae he observed were rotating, in the sense of unwinding.[26] When these results were published, they were readily accepted by the majority of astronomers because they fitted the then prevailing theoretical considerations, and also because they agreed with the data accumulated in the period 1914–16.[27] Moreover, some astronomers repeated van Maanen's observations and confirmed them. Any criticism of van Maanen's data at that time was silenced. So, when the astronomer Kurt Lundmark wrote to Harlow Shapley of Harvard University about his doubts concerning the correctness of the data, Shapley replied to the effect that people who lived in glass houses should refrain from throwing stones, and furthermore, that one could also find flaws in Lundmark's conclusions in his own important papers about the distances of globular clusters. Lundmark did not pursue the matter further.

Eventually doubts arose when some astronomers, in private correspondence, claimed to have found discrepancies between the rotation data and additional evidence. Van Maanen's measurements of rotation of nebulae were finally disproved by Edwin Hubble, also of Mount Wilson Observatory, in 1924. Hubble showed that spiral nebulae were in fact galaxies at immense distances from Earth. Initially Hubble tried to attribute the discrepancies between his and van Maanen's data to random error, but it soon became obvious that van Maanen's results were 'obtained because he had read his expectations into his data'. Hubble criticized van Maanen's results in internal colloquia and later in a public statement.[28]

Hubble made his criticisms despite attempts to dissuade him from doing so by the administration of the Observatory. It was felt that disputes between scientists of the same institution should be settled 'in house' and that the publication of criticism would disrupt morale within the institution.

A third astronomer from the Mount Wilson Observatory whose data were later criticized as not feasible was George Eilery Hale (director of the Observatory at the time). He claimed that on the basis of his observations one could assume that the sun had a magnetic field. His argument ran thus. At the temperature of the sun many elements were ionized and emitted identifiable spectral lines. When a beam of white light is passed through a triangular prism of glass, the beam spreads into a band of rainbow colours. In such a coloured 'spectrum' of sunlight there appear dark lines (spectral lines) which had been identified as being due to the presence of ionized elements in the atmosphere of the sun. These lines thus identify the elements present in celestial bodies. It was known that such spectral lines could be split by strong magnetic fields (Zeeman's effect). Hale confirmed van Maanen's previous observations in his 1914 photographs of the sun that Zeeman's splitting indeed occurred in the solar spectra. An English astronomer who measured the images in Hale's photographic plates, however, obtained contrary results, and published his discrepant views. Other astronomers who tried to repeat Hale's observations failed to find the splitting phenomenon, and came to the rather lame conclusion that during the time that had elapsed since Hale's measurements the magnetic field of the sun had declined in strength. When Hale's plates were re-examined once more, it was established that they did not indicate the presence of any significant magnetic field in the sun at any latitude. The conclusion published by Stenflo[29] was that Hale's results were systematic errors due to personal bias.

These cases confirm the hypothesis entertained by Rosenthal[19] on the existence of experimenter bias, i.e. that the expectations of observers or experimenters lead to a systematic error in their observations. While in medicine and psychology one uses double blind procedures which help to offset the effect of experimenter bias, the situation is different in astronomy because of the difficulties in repeating the experiments and also because often the observations are made at the very limit of accuracy and sensitivity of the instruments available (see also chapter 3). According to Hetherington, 'the warping of judgement by knowledge, the influence of observational reports on preconceived opinion, is inevitable'.[24] Hetherington also felt that there was something wrong in the attempt to explain discrepant data as being due merely to an experimental error. He felt that van Maanen's findings that all seven nebulae he observed were unwinding could not have been attributed to random error, because random error would produce results interpreted as either indicating winding up or unwinding of the nebulae.

A doctoral thesis from Boston (R. Hart, 1983) that attempted to find other sources of error in van Maanen's findings, however, eventually concluded that the only explanation of them was an error based on bias. The conclusion Hetherington entertains is that although none of the Mount Wilson Observatory cases could be considered as instances of conscious fraud, they were nevertheless an indicator of personal bias, or at the very least, a failure in scientific objectivity.

3

Error or Self-Deception?

Apart from the 'experimenter effect', any scientist will almost inevitably make an error at some point in his or her investigations because of the vagaries of scientific thought, and may draw the wrong conclusions from the results of an experiment. Though such errors, when detected, may be disappointing to the individual in question they are nothing to be ashamed of; they do not affect the course of scientific thought and may be easily corrected (or forgotten). When, however, the original faulty observation or conclusion is made by a scientist of authority in the field, and it fits the established ways of thinking of the scientific community about the phenomenon in question, it may start an avalanche of *bona fide* experimentation, where the same error is repeated and thus non-existent phenomena become confirmed.

THE N-RAYS THAT NEVER WERE

A classic example of this kind of occurrence is provided by the discovery of 'N-rays' by a distinguished French physicist, René Blondlot, Professor of Science at the University of Nancy during the years 1901–4.[30] Blondlot experimented with X-rays, a few years after their discovery by Roentgen. During these experiments, Blondlot observed a new type of radiation, which he named 'N-ray' radiation as a tribute to Nancy, the city where he worked. These rays were emitted by a very hot platinum wire enclosed in an iron tube. Having passed through a thin aluminium window, they were directed at either a gas flame or a faintly illuminated calcium sulphide screen, analogous to the TV screen. The N-rays would increase the luminosity of the flame or of the white screen. Blondlot also claimed that these rays could be stored. A brick wrapped in black paper or aluminium foil and exposed to the sun would store the N-ray energy which could then be detected in the laboratory when the brick was brought close to the white surface.*

Blondlot's experiments caught the attention of renowned and competent physicists such as Charpentier, Becquerel, Broca, Zimmern and others. They repeated Blondlot's experiments, with apparent success and confirmed his findings.

It must be remembered that at that time there was great enthusiasm for new radiations, ushered in by the discovery of X-rays and of radium. The time was ripe for other discoveries of this type. Within four years of the discovery, tens of papers confirming the existence and the properties of N-rays were published in 'respectable' journals.

In 1904 an American physicist, R. W. Wood, came to France to visit Blondlot's laboratory and observe his experiments. The experiment Blondlot was about to demonstrate to Wood involved bending and measuring the bend of N-rays. The rays were supposed to emerge through a 2 mm slit, and be bent by an aluminium prism (in the same way as light is bent by a glass prism), before falling on to the target screen to be measured. After the first demonstration, having asked for the experiment to be repeated, Wood surreptitiously pocketed the aluminium prism without Blondlot noticing. In spite of this, the results of the second experiment were exactly the same as with the prism in place. Wood published the story in *Nature* in 1904, and also in *Physikalische Zeitschrift*. In 1909 Blondlot retired from his professorship. It seems to me that it was Blondlot's enthusiasm and belief in his discovery that had led him astray. The phenomenon of N-rays depended on a threshold perception of faint luminosity, and once some preconceived physical calculations were available for the properties of the new rays, their observation could easily be steered by this foreknowledge. Blondlot believed he made a discovery. In his own words, related in the Proceedings of the *Académie des Sciences* of 23 March 1903, he stated:

> Previously I had ascribed polarization to X-rays when, in fact, it was produced by the new rays; this error was unavoidable before the

---

* The material on N-rays and on scotophobin has been taken from Kohn, A. 1978: Errors, fallacies or deception? *Perspectives in Biology and Medicine*, 21: 420, by permission of the University of Chicago Press. See also, for more detail, Nye, M. J. 1980: N-rays: an episode in the history and psychology of science. *Historical Studies in the Physical Sciences*, 11: 125.

study of refraction effects had been completed. Only after that study did I become convinced that I was not tackling X-rays but an entirely new type of radiation.

It is not clear whether indeed this belief biased Blondlot in his further studies, and the excitement in the scientific community led other scientists as well to confirm Blondlot's findings, or, as it had been suggested by Lucien Cuénot (and quoted by Rostand)[30]:

> The whole discovery of N-rays might have been initiated by an overzealous laboratory assistant who tried to make himself indispensable to his professor . . . assistants are not usually given to scrupulous love of truth and have little aversion to rigging experiments; they are quite ready to flatter their superiors by presenting them with results that agree with their *a priori* notions . . .

THE DAVIS AND BARNES EFFECT

Another phenomenon that, like the discovery of N-rays, was based on threshold visual discrimination and illusory results was the so-called 'Davis and Barnes effect'. It was based on observation of scintillations on a zinc sulphide screen produced by a beam of alpha particles emitted by polonium in a vacuum tube. In the absence of a magnetic field, the alpha particles would proceed in a straight line and be collected on one zinc sulphide screen (Figure 1); when a magnetic field was

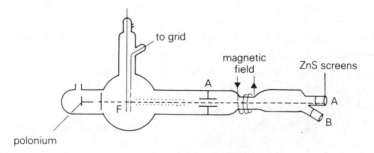

Figure 1. The Davis–Barnes effect. Alpha particles (polonium), emitted in an evacuated glass vessel, pass through a grid (G) together with electrons produced by a heated element (F), and are accelerated by an anode. They continue through a magnetic field which can deflect the beam from a ZnS screen (A) to another screen (B), and there produce scintillations.

applied to the tube, the particles would be deflected on to another screen within the tube. There was also a hot cathode electron emitter in the tube, which emitted a stream of electrons moving with the alpha particles. These electrons could be accelerated by the application of a suitable voltage to a grid interposed in their path. Davis–Barnes found that when the velocity of the electrons was adjusted to that of the alpha particles, the electrons were 'captured' by the alpha particles, endowing them with a partial negative charge. The alpha particles with the captured electrons would now be deflected in the magnetic field to another part of the zinc sulphide screen. Simple counting of the change in the number of scintillations would indicate whether the electrons were captured or not.

The extraordinary thing was that it seemed the electrons would combine with the alpha particles at 590 volts (applied to the grid), as well as at other calculated voltages provided that the velocity differences between the electrons and the alpha particles corresponded to some theoretically calculated velocities. The sensitivity of the Davis and Barnes system was such that even a change of one-hundredth of a volt would affect the results.

Irving Langmuir, a famous physicist who heard Davis describe these results at a seminar at Knoll Research Laboratory in Schenectady, NY, in 1929,[31] expressed his doubts about the whole experimental set-up. He went to Davis's laboratory at Columbia University, accompanied by Drs Whitney and Hewlett, to observe an actual experiment. The experiment was performed in a darkened room. The scintillations were counted by Langmuir, Hewlett or Barnes; an assistant, Hull, was sitting at a panel adjusting the voltages. When Langmuir arranged the situation so that Barnes had no clue as to the voltage settings adjusted by Hull, half of his readings were right and half wrong. Langmuir then wrote a 22-page letter to Barnes showing that the experimental approach was wrong and that he (Barnes) was counting 'hallucinations', a situations not uncommon among people who work with scintillations!

Nevertheless, eight months after Langmuir's visit, Barnes and Davis published a paper in *Physical Review* on their effect.[32] The same year Webster,[33] using a Geiger counter in place of the scintillation screen, explained the results recorded by Barnes and Davis as probability error. (A Geiger counter is an instrument that objectively records any ionizing radiation that enters it through its 'window'.) Webster's findings eventually led Barnes and Davis to make the following statement in *Physical Review*:

The results reported in the earlier paper depended upon observations made by counting scintillations visually. The scintillations produced by alpha particles on zinc sulfide screen are a threshold phenomenon. It is possible that the number of counts may be influenced by external suggestions or autosuggestion of the observer.[34]

THE ALLISON EFFECT

A third example of an illusionary physical phenomenon, reported in the same decade as the Davis and Barnes effect, concerned the 'Allison effect'. Allison's observation was based on the so-called 'Faraday effect', in which a beam of polarized light passing through a liquid placed in a magnetic field is rotated. In 1927 J. W. Beams and F. Allison found that there was a time lag between the removal of the magnetic field and the disappearance of the Faraday effect. They published their findings in the journal *Physical Review* (1927, 29:161) They constructed an electro-optical apparatus (Figure 2) in which the glass cell containing the tested liquid was surrounded by a coil which would induce a magnetic field in the liquid when a current was passed through it. The rotation of the polarized light passing through that solution was measured with the aid of two Nicol prisms at the two sides of the cell. The electrical circuit was arranged so that the same current that induced the magnetic field produced a spark, generating the light to be polarized. The length of the wire carrying the current could be varied by moving a trolley which at the same time

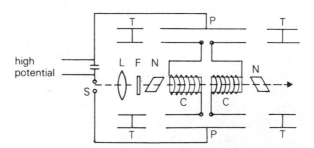

Figure 2. Allison's magneto-optic set-up. L, converging lens; F, Wratten filter; C, movable cells containing the liquid to be tested. The cells are surrounded by a coil producing the magnetic field. N, polarizing Nicols; S, spark gap; sliding trolleys. Arrow shows the light path (adapted from Allison, F. and Murphy, F. J. 1930: *Journal of the American Chemical Society*, 52: 3796).

also changed the optical distance between the Nicol prisms. With a particular liquid the system would be adjusted to obtain a minimum light intensity, and then the cell moved so as to obtain the same minimum after removing the magnetic field. The distance the cell moved divided by the velocity of the light would give the time lag. The accuracy of the system was such that a time delay of 10^{-10} seconds could be measured.

Beams and Allison found that different liquids produced different time lags. This delay was also dependent on the wavelength of the light used.[35] Allison next established that in salt solutions each chemical compound, regardless of the presence of other substances, produced a characteristic minimum light intensity, even at extremely low concentrations (one part in 100 billion). This observation led to the determination of the existence of various isotopes of metals[36-39] and the discovery of a new element, virginium. Latimer and Young,[40] who discovered tritium by this method, stated later that they could never repeat the experiment.

In 1932 McGhee handed Allison 12 unknowns to be tested. Allison identified all of them correctly within three hours.[41]

In 1935 Jeffesen and Bell made an objective study of the Allison effect. They asked 150 observers to make measurements by the Allison method of coded substances. Some 4000 readings were made during a year and analysed. Jeffesen and Bell reached the conclusion: 'Minima in fair agreement with Allison's result have been obtained with the usual set-up, but just as good agreement had been obtained with wrong solutions in cells, or none at all. This indicates that the minima are not a function of the chemical solution used.'[42] The results were, they concluded, a probability effect.

The Jeffesen and Bell paper seems to be the last on the Allison effect, after several hundreds of papers had been published on the subject. It is worth noting again that, as in previous cases, the use of a complicated electro-optical apparatus was associated with *subjective* estimates, in this case of minima of light, a situation which we have seen lead many experimenters to 'probability' results.

THE MYTH OF MITOGENETIC RAYS

One epidemic of scientific delusion, resulting in some 500 publications in the 1920s and 1930s, involved 'mitogenetic rays', a phenomenon linking physics with biology. Their existence was first reported by Alexander Gurwitch.[43,44] They were defined as ultra-violet (UV) rays, of a wavelength below 2500 A, which were filtered out by glass but not by quartz, and were emitted by animal and plant cells when they were dividing. This radiation had very low energy, estimated to be about 10–100 quanta per cubic centimetre per second.

In Gurwitch's original experiment, an onion root was permitted to grow in a narrow glass tube. Its emerging tip was exposed to another tip of a growing onion root approaching the first at right-angles from a distance ranging between 0.5 and 12 cm. A quartz or glass sheet was interposed between the two tips. At the end of the exposure period, which lasted from 10 seconds to 60 minutes, the tip of the target root was sectioned and the number of dividing cell nuclei in the near and far halves of the section counted. When the two root tips were separated by quartz, the number of cell divisions in the half of the target root nearer to the radiating tip was significantly larger than in the further half. When the radiation barrier was glass, there was no difference in the number of cell divisions in both halves.

In later experiments Alexander Gurwitch replaced the onion roots by yeast and bacteria. He grew the emitting (sender) yeast cells in a liquid suspension in a horizontal quartz tube; under it, separated by glass or by quartz sheets, there were two vessels containing bacterial suspensions (Figure 3). Wolff and Ras[45] found that the number of bacteria in the vessel separated from the sender by quartz increased at a more rapid rate than in that screened by glass.

It was experiments of this type that demonstrated the existence of so-called mitogenetic rays by their biological effects. The research later extended to other biological fields. There were reports that the blood of healthy children, but not of those suffering from vitamin D deficiency, emitted mitogenetic rays; that implantation of cancer cells into an animal would arrest the radiation from it, etc.[46] The existence of the phenomenon was supported and contested by an equal number of scientists.

Since, after all, the causative agent was supposed to be UV radiation, its existence should have been demonstrable by purely physical means. These rays should have been detected by photographic plates, photoelectric cells, or by changes in some radiation-sensitive chemical reaction. Many such measurements proved negative, but it was claimed that the energy of the mitogenetic rays was too small to cause blackening of photographic emulsion on short exposure. Nevertheless, even exposure of *months* to

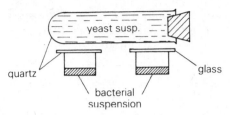

Figure 3. Experimental arrangement to measure the presence of mitogenetic rays. The 'rays' emitted by yeast cells growing in a quartz tube (*top*) affect the multiplication of bacteria in the quartz vessel (*left*) but not in the glass vessel (*right*) (adapted from Hollaender, A and Claus, W. D. 1935: *Journal of the Optical Society of America*, 23: 270).

daily changed senders did not affect ultrasensitive photographic plates.[17, 48] Moreover photoelectric cells, sensitive to less than one-third of the light intensity of the mitogenetic rays required to evoke the biological effect, were not able to detect their existence.

The whole research programme on mitogenetic rays has been critically reviewed by Hollaender and Claus.[46] They concluded that the field had been in a chaotic state, that a large number of the 500 publications on the subject were either erroneous or contradictory, and that 'more convincing clear cut evidence for the existence of the phenomenon must be established before it can be accepted as real'. In an editorial in *Nature* (7 February 1931, p. 214) B. P. Tokin makes it clear that this new experimental field had failed to produce any evidence of 'mitogenetic rays'.

The demise of mitogenetic rays, however, which would have been expected after Hollaender's review, was not complete. In Poland, Rylska[49] published a paper supporting the phenomenon. Although in 1960 Moiseva[50] came to the conclusion that mitogenetic rays had been fully discredited, and recommended that work in this field be discontinued, there yet appeared six years later papers by Anna Gurwitch (the daughter of the inventor of the mitogenetic rays) describing the emission of these rays from the myocardium of frogs, rabbits and cats.[51, 52]

The mitogenetic rays, as such, do not exist. Today we may perhaps name them 'mythogenetic' rays. Nevertheless, Hoefert[53] has established that living cells emit visible light (chemiluminescence), but not any short wave UV radiation.

In this, as in the other cases we have looked at, there was no dishonesty involved. The researchers obtained false or erroneous results by being led astray by subjective effects, wishful thinking or threshold observations. Langmuir[31] calls this 'pathological science' which is characterized by the following symptoms:

(1) the maximum effect produced by the causative agent is barely detectable, and there is no correlation between the magnitude of the effect and the intensity of the cause: the observations are near the threshold of visibility;
(2) the data accumulated in such a sct arc extremely accurate and serve as a foundation of rather fantastic theories;
(3) at the height of the publication of the phenomenon usually half the scientific community are supporters, claiming success with the experiments, and half critics, who cannot reproduce the phenomenon.

POLYWATER - AN ANOMALY EXPLAINED

In 1962 a Soviet scientist, N. N. Fedyakin, reported the discovery of a water-like liquid, formed during condensation of water vapour in quartz capillaries. He believed this substance to be an anomalous form of water,[54] and based his conclusions on the peculiar properties of this water. It had a density some 40 per cent higher than that of water, froze at $-40\,°C$, and had the general consistency of petroleum jelly.

It is important to describe here the experimental procedure that led to the production of polywater (as this new form of water later become known). Fedyakin placed ordinary distilled water in a small container within an enclosure that could be evacuated; he then allowed the water vapour to condense inside quartz capillaries suspended above the evaporating dish. After a few days or weeks, he took out the quartz capillaries and examined their contents, pushing them out from the capillaries with a wire. They contained the new form of water, which evaporated and boiled at temperatures different from those of ordinary water.

Fedyakin later joined the laboratory of Boris Deryaguin, a surface chemist the Institute of Physical Chemistry in Moscow. There a whole group of scientists participated in the study of the anomalous water and published their findings in some 16 papers. By this time the experimental design had become more sophisticated. The evaporation and condensation of water was arranged so that the ordinary water enclosed in one quartz capillary has heated to 160 °C, make it boil and evaporate, and next to it was placed an empty quartz capillary where the water vapour condensed. In a still more advanced design the water was placed at one end of a sealed quartz capillary. The central part of this capillary was wrapped with a heating coil which was heated to 500 °C; at the other end, the capillary was cooled to freezing point (Figure 4). In this arrangement only the water vapour could cross the heat barrier to be condensed in the cooled part, and there was no possibility of the water creeping across the barrier. The water condensed in the cooled part was then examined: it was found to have the peculiar properties of polywater, a supposed polymer of water.[55]

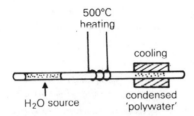

Figure 4. The apparatus used in the preparation of polywater (From Kohn, A. 1978: *Perspectives in Biology and Medicine*, 21: 420).

In 1966 Deryaguin visited England and lectured about the anomalous water to the Faraday Society in Nottingham. The idea of polywater appealed to some British scientists and Deryaguin's experiments were first repeated by Bernal at Birkbeck College, London, and later by other scientists, all confirming the findings of the Russian scientists.

By 1968 the scene switched to the USA, where Lippincott and his colleagues published an extensive paper on their experiments with polywater.[56] On the basis of infrared and argon laser Raman scattering spectra, they firmly concluded a new form of water was created in a polymerized form and they gave it the name 'polywater' (the Russians had until then called it 'anomalous water'). Lippincott went even further and proposed how the atoms of polywater were organized in space in their lattice so as to account for the density and viscosity of the anomalous water.[57] The theoretical chemist Allen then developed a theory to account for the data and the model of Lippincott and Linnett (Allen and Kollman, *Science*, 1971, 167:1443).

Numerous scientists joined the research on polywater. In the resulting avalanche of papers (reviewed by Allen[58]), a rare word of caution was found in a letter in *Science* from J. H. Hildebrand of the University of California.[59] He wrote:

> . . . we are sceptical about the contents of a container whose label bears a novel name but no clear description of its contents; . . . we are suspicious of the nature of an allegedly pure liquid that can be prepared only by a certain person in such a strange way. I think that a chemist who feels curious about what is in those glass capillaries would have more success if he assumes that he is dealing with a system of two components.

In all these studies the main problem was that the quantities that could be obtained by the capillary method were minuscule (microgrammes), and this made the detailed analysis of the product difficult.

Some hundreds of papers had appeared in the literature by 1971, when the whole edifice began to collapse. Already in 1969 a Soviet mass spectroscopist, Talrose, had claimed that Deryaguin's samples of polywater were contaminated with fatty substances. In Lippincott's samples traces of sodium and carbonate were detected. These findings, it seemed, called for the use of more sophisticated analytical methods.[60-62] Finally, it was established that the 'polywater' condensing in narrow quartz tubes was a solution of the element silicon in water contaminated with 20–60 per cent sodium, 3 per cent calcium and potassium, 15 per cent chlorides and sulphides and some lactates and phospholipids. Rousseau and Porto[60] calculated that the source of impurities could not have been the quartz capillaries themselves; they attributed the contamination to salts that some researchers were using to reduce the vapour pressure of the evaporating water, or to the grease used to tighten the seals. A further possibility was that in some experiments the water diffused along the walls of the capillaries, leaching out the impurities. They summed up their findings by saying that the existence of polywater was unlikely.

Evidence accumulated that polywater was a colloidal silica gel contaminated with carbon dioxide from the atmosphere as well as with other molecules. Kurtin *et al.*[63] measured the dielectric properties of polywater and found it to be a two-phase system consisting of finely divided particulate matter suspended in ordinary water. The presence of such particulate matter in polywater was indeed confirmed by scanning electron microscopy. It thus became quite clear that polywater did not exist, and that all the results that had led scientists to conclude that it did, were erroneous. The results could be easily explained by contamination occurring during the handling of the silicon (quartz) capillaries.

In 1971 Allen presented new calculations that discredited his previous theoretical model.[64] He also said:

> We have reversed our conclusion from initial belief in the existence of a new allotype of water to our present proposal of silicate-plus-anion explanation. This will cause many chemists to laugh and some to discount the efficacy of theoretical chemistry. It is certainly true that for the complex processes on the microscopic level, theory has not achieved the striking results it has enjoyed in elucidating the basic force laws in physics, but the increasing volume and diversity of chemical knowledge demands the simplification and organizational powers of theory more than before.[65]

Was the polywater fallacy inevitable? Could the first observations and data have warned the researchers that polywater was an artefact? Why did so many scientists follow the false lead, and repeat experiments and measurements that were not well controlled? Allen had this to say on the subject:

> My own involvement in the polywater phenomenon greatly strengthened my belief in the scientific method; I was greatly stimulated by both the human and scientific experiences, and the attention focussed on my first paper on the subject gave me a more convincing sense of having made a true contribution to science progress than I have felt for some other conventionally successful research projects.[66]

In his book on polywater, published in 1980, Franks[67] thought that possible criticism of the discovery of polywater by Fedyakin was prevented by the influential scientist Deryaguin, when he himself claimed priority in this discovery. Once polywater became a popular subject, the people involved in the research liked to attract the attention of the press (even if they denied it), and of their peers at seminars and meetings. Even J. D. Bernal referred to the polywater discovery as 'the most important physical chemical discovery of the century', but he ridiculed the idea held by a number of scientists at the time that polywater escaping from a laboratory could autocatalytically polymerize all of the world's water, ending life on earth!

Ten years of excitement over polywater abated, presumably making the scientists involved more wise and more cautious. It should be remembered that the rise and fall of polywater was the product of genuine experimentation. Simply because of its novelty and the enthusiasm for it some warning signals were not heeded. In the end, however, the truth was uncovered and the matter satisfactorily explained without causing undue damage to those involved. Nevertheless, one wonders how, in spite of all the sophisticated analytical techniques available, the contaminants causing the water to change its properties were not detected earlier.

SCOTOPHOBIN - FACT OR FICTION?

The idea that learning is connected with chemical changes in the nerve cells in the central nervous system, and particularly in the brain, has been attractive for many decades. Memories and learned patterns of behaviour must be registered somehow in the brain, and many biologists have favoured the idea that they are vested in macromolecules – molecules of proteins or nucleic acids – which are so complex in structure that they could carry such information in coded form.

James McConnell of the University of Michigan demonstrated that flatworms (Planaria) could be taught behavioural patterns in a maze by coupling light stimuli with electric shocks. He also showed that the learned patterns could be transferred to naive flatworms either by feeding them with pieces of 'learned' worms or by injecting them with RNA extracts from trained Planaria. Flatworms are about one inch long; they live in the bottom of ponds, streams and rivers throughout the world. They have a true brain and a synaptic type of nervous system. McConnell interpreted his experiments as indicating that the injected worms acquired stimulus-response patterns that the original donors had acquired only by hard experience.

Though such experiments were successfully repeated by a number of scientists, there were also groups who did not succeed in effecting a similar transfer of learning in Planaria, mice and rats. In 1966, 22 scientists from different laboratories published a joint paper in *Science*[68] declaring that they had tried, but failed, to transfer learned patterns (memory) from one group of animals to another. In 1970 a comparative review of some 400 studies on transfer indicated that only half of such experiments were successful. (Does this not strike a chord reminiscent of Langmuir's pathological science?)

Against this background came the work of Georges Ungar of the Baylor College of Medicine in Houston. In 1968 he reported that he had trained rats to avoid darkness. He used a training box designed by Gay and Raphaelson of the University of Michigan. It had a row of three chambers connected by open doors. The centre and one of the end chambers were white; the other end chamber was black. Rats, being night-adapted animals, would prefer to go into the black chamber when put in the middle between the black and the white enclosures. Ungar administered an electric shock to the rats that moved into the black chamber and thus taught them to avoid the dark compartment. The trained rats were then killed and their brains extracted. Ungar injected naive rats with these brain extracts and observed that they acquired the fear of darkness (hence the name scotophobin).[69] During the following two years Ungar collected brain material from some 4000(!) rats that had been trained to fear darkness. From these brains he isolated a polypeptide which he named scotophobin. In 1972, Ungar, Desiderio and Parr[70] identified the scotophobin as a polypeptide made up of 15 amino acids. The sequence was:

ser-asp-asn-asn-gln-gln-gly-lys-ser-ala-gln-gln-gly-gly-tyr-NH2

The remarkable aspect of the publication of Ungar's paper describing the structure of scotophobin was an accompanying article (almost twice as long) entitled: 'Comment on the chemistry of scotophobin' written by W. Stewart of the National Institutes of Health.[71] Stewart was the referee of Ungar's paper. It seems exceptional that an original article and a review of it should have been published side by side in the same journal. The editors of *Nature* added an introduction stating that 'these comments represent Stewart's *remaining* reservations after a full consultation with the authors'.

Stewart's article was very critical of Ungar's methods, results and conclusions. He wrote: 'They intended their article as a brief letter to *Nature*, after their findings have (already) been presented in one note, five abstracts, three symposia and eight review articles, a total of 100 pages . . .' but in none of these, he said, were enough experimental details given to allow duplication of experiments. Stewart also noted that no unequivocal relationship existed between the dark avoidance and chemical transfer of learning. He based this conclusion on two statements made by Ungar, one published in 1968[69] and the other in 1970.[72] The first stated: 'The experiments reported here are easily and rapidly reproducible and yield unequivocal results which clearly demonstrate the possibility of a purely chemical transfer of some types of acquired information'; and the other, of 1970, reads: 'Not all our experiments were successful and occasional failures have been experienced by other workers in this area. To succeed, too many conditions, most of them unknown, have to be fulfilled and if even one of them is neglected, the whole experiment may fail'.

By selecting these quotations, Stewart called into question the reproducibility of Ungar's experiments. Stewart's other criticism concerned the structure of scotophobin, which Ungar identified as a definite peptide made up of 15 amino acids. Stewart remarks: 'By publishing only the data consistent with their [Ungar *et al.*'s] interpretation, and by omitting all the rest, the authors leave little room for independent evaluation of their reasoning'. He continues: 'A clear implication of these calculations about the sequence of amino acids in scotophobin is that there are not merely eight structures as the authors imply, but rather several thousands which are consistent with both chemical and mass spectrometric data.' There is no need to add to this criticism, which, in simple words, said that the existence of scotophobin, as a definite polypeptide with its particular biological properties, was extremely doubtful.

During the years 1972–5 a number of groups in various countries experimented with scotophobin, either using the original material extracted and purified by Ungar and supplied by him, or a synthetic material made according to Ungar's data.[73] In 1973 another group of scientists[74, 75] found that scotophobin was inactive in rats tested in the Gay and Raphaelson apparatus, as well as in an open field test. They thought that their failure might be due to some other unpleasant stimuli to which the animals might have been unintentionally exposed. They suggested that scotophobin, being, according to its structure, similar to ACTH and vasopressin, both hormones that elicit fear and raise blood pressure, might have had a stress-producing effect; it would thus increase the motor activity of the animals and shorten the time spent by the hyperactive rats in a dark box. A study by Miller, Small and Berk[76] pointed to a possibility that the rats used as donors of scotophobin did not actually learn to avoid darkness, but avoided other apparatus clues. Indeed stress effects during the training of animals receiving electric shocks have already been described.[77]

Ungar died in 1977, and Dr Parr, who was responsible for the synthesis of scotophobin, returned to the Federal Republic of Germany, where he now works for an industrial firm.

By 1978 many doubts were arising concerning Ungar's studies. A sample of scotophobin, prepared by Parr, was analysed in 1979 by the Knauer Science Company in Berlin and found to contain at least 40 different components. The National Institute of Mental Health, which had financially supported Ungar's studies, began to question the authenticity of the phenomenon of dark avoidance. It was agreed that the Baylor chemists would synthesize scotophobin according to the formula Ungar had defined as representing the active material, and that this synthetic material would be sent to interested laboratories for re-testing. Unfortunately Ungar's chemists at Baylor 'were never able to synthesize and deliver any material which had been found to have scotophobin activity' (Private Communication, Dr Usdin, NIMH).

Professor Weinstein of the Department of Chemistry at Washington University obtained from NIMH a few milligrammes of scotophobin: 'it was found to be nothing but a complete mixture of peptides having no activity' (which scotophobin should have) (Private Communication, Weinstein).

The Baylor chemists concluded that the structure as previously determined at Baylor was in error, and proposed that NIMH should fund their research to re-determine the structure. At this stage the support for the project was terminated.

The question of whether scotophobin exists as a defined peptide inducing fear of darkness in rats can be answered in the negative. In a book on *Endogenous Peptides and Learning and Memory Processes* published in 1981[78] scotophobin is not mentioned at all. The interest in scotophobin has nevertheless not died completely, in spite of the adverse publicity. In 1976 George Maheryshyn, a graduate student of Dr Helene N. Gutman of the University of Illinois (Chicago Circle), described in his unpublished master's thesis the isolation from the brains of 2000 goldfish that had been trained to avoid darkness of a polypeptide made of 15 amino acids said to be goldfish scotophobin. It differed from the alleged rat scotophobin by one amino acid at the end of the molecule. During the eight years that have passed since, nothing more has been heard of this research.

The questions still go unanswered: how did this alleged scotophobin keep many scientists in a number of countries alert and interested for ten years, and why was so much public money spent on the research?

My guess is that the excitement was due to the fascination with the problem of memory. There are many laboratories throughout the world studying it by various methods. Progress is slow, but already there are indications that connection between memory storage and genetically coded proteins is being established. Every year new discoveries of peptides active in the nervous system are described. A scotophobin-like peptide may exist, but it yet remains to be clarified what are its true role and function. On this tortuous road to truth there are always those who stumble and are left behind.

A last example concludes this chapter on error and self-deception. In 1972 Alessandro Bozzini of the Centre for Nuclear Energy in Italy found that some strains of wheat contained protein at a very high concentration. These strains lacked a piece on the short arm of the chromosome 2A. Bozzini interpreted this finding with a hypothesis that the lacking piece was the seat of a gene coding for an inhibitor of protein synthesis in the seeds of wheat. In the absence of this gene the protein synthesizing machinery is not inhibited and more protein is produced than in plants with an intact chromosome 2A.

Bozzini's findings and interpretation stimulated a burst of research activity supported by the International Atomic Energy Commission – since ionizing radiation could induce similar mutations in wheat as those observed by Bozzini. Within a few years, however, it became apparent that Bozzini's result had been due to smaller sized seeds in the strain lacking the piece of the chromosome. Production of the same amount of protein in a seed of smaller volume results in higher concentration (amount/volume) of protein. We encounter in this case a *bona fide* error in the interpretation of results.

4

In the Shadow of Doubt

We highly revere the famous scientists who were responsible for the progress of science through the ages. Among them, however, we find a small but significant number who may have 'cut corners'; these scientists seem to have presented evidence because their intuition told them something was true. I shall examine here the doubts that have arisen concerning the work of Ptolemy, Newton and Mendel.

PTOLEMY UNDER SCRUTINY

At the beginning of the nineteenth century the French astronomer Delambre accused Claudius Ptolemy, the famous astronomer of ancient times, of not actually having observed the celestial positions at around AD 135, positions which Ptolemy describes in his book *Almagest*. Delambre's accusation was later supported by an American astronomer, R. R. Newton, of Johns Hopkins University. Delambre and Newton claimed that Ptolemy's alleged observations of the equinox in Alexandria were merely extrapolations from data of Hipparchus (who some 200 years earlier had discovered the Earth precession). They thus attributed Ptolemy's work to forgery.

More recently this view was repudiated by Neugebauer[79] in his book on the history of ancient astronomy. In Neugebauer's view, Ptolemy was the first astronomer to note the constancy of the tropical year, that is the time required for the sun to return to the same position in respect to the equator. Neugebauer, as well as Owen Gingerich,[80] think it unlikely that Ptolemy's success rested on fabricated observations, but that 'in those ancient days, before error theory was understood, selected observations were adjusted for pedagogic purposes and recorded in *Almagest* in close agreement to theory . . .'.

The difficulties in obtaining reliable data in the time of Ptolemy were enormous, and it is only to the credit of the ancient astronomers that they succeeded in building a theoretic structure of astronomy. The cinematic theories of *Almagest*, backed up by better observational techniques, were essential to Newton's celestial mechanisms.

DID NEWTON FUDGE HIS DATA?

Newton, one of the fathers of modern science, continued the search for mathematical simplicity in physical phenomena, a search that began with Copernicus, Kepler and Galileo. In book III of his *Philosophiae Naturalis Principia Mathematica* published in 1697, Newton attempted to reduce the context of his philosophical principles of the System of the World to popular mathematical treatment that would be understood by those who had mastered the first two books. Till then the Pythagorean tradition represented by Kepler and Galileo was based on geometrical description of natural phenomena (movement of planets and of falling bodies), while the mechanical tradition, represented by Descartes, sought mechanical causes of these phenomena. It was Newton's genius that united these two concepts by first bringing into science the concept of action as force at a distance.

Kepler's law of planetary motion did not take into account the effects of Earth, and Galileo's calculations of the movement of planets were only ideal representations of real events. Newton was the first to take into account in his *Principia* the perturbations which differed from the ideal concepts on which the science of the time was based. He demonstrated that these perturbations could be dealt with mathematically, and he showed that theoretical calculations based on his principles would lead to results that were in accord with the experimentally measured data. Unfortunately, in striving to demonstrate how well the material events corresponded to the mathematical calculations, Newton presented his data with a degree of precision that was practically impossible to obtain in his time.

On the basis of his principles, Newton calculated the correlation between the distance to the moon and the gravity constant, the velocity of sound and the precession of the equinox.

The law of universal gravitation was derived from the correlation between the acceleration of falling bodies due to gravity on Earth, and the acceleration of the moon which moves around the Earth as if it were trying to fall. From Kepler's data Newton knew that such centripetal attraction of a planet to the sun inversely varied with the square of the distance from the sun. Newton concluded that the attraction that held the moon in its orbit around the Earth was correlated with gravity as measured by the acceleration of falling objects on Earth. He thus took into consideration the sun's effects on the moon's orbit, the moon's centripetal acceleration, the ovoid shape of the Earth (affecting the rotation of Earth around its own axis) and came to the mathematical conclusion that ½ *g* should equal 15 feet, 1 inch and 1.5 lines. The precision of this mathematically derived number was better that 1 part in 30,000; this precision exceeds that which was feasible in Newton's times, when, for instance, Boyle's measurements, to confirm his law were accurate to within only 1 in 100.

Newton's derivation of the velocity of sound was an achievement of a genius. Newton considered the length of a wave in water (the distance from crest to crest) and its velocity of propagation. He then proceeded to calculate how sound would behave if it were indeed a wave. He needed to know the ratio of the density of air to that of water, a value which was not known. He assumed it to be 1:850. Using Boyle's law to translate the data for density of air and of compression waves in the air, he then calculated that the velocity of sound should be 979 feet per second. In order to confirm this calculation Newton performed the famous experiment in Neville's Court in Trinity College, Cambridge, in which he measured the time of return of an echo. He used pendulums of different lengths to measure the time of sending out the sound and its return to the point of origin. His measurements provided him with a range of values of 920–1085 feet per second. Thus, his calculated value of 979 feet per second was well within that range. This value, however, was about 20 per cent lower than the experimentally derived estimates of the velocity of sound as measured in 1708 by Derham, as well as by Halley and Flamsteed of the Royal Observatory. Having learned these experimental values, Newton decided to make a further assumption that would take care of the 20 per cent discrepancy between the experimental data and his calculation. He assumed that air waves did not involve geometrical points, but real air particles (he called that 'crassitude' of air); next he assumed that the air contained water vapour at the ratio of 1:10. When these assumptions were introduced into Newton's formulae, each contributed to a 10 per cent increase and thus the now calculated velocity of sound was 1142 feet per second, a figure which coincided with the experimentally measured number to within one part in a thousand.

How did Newton arrive at the numerical values for his assumptions? He had no factual basis for them; he did not have data to show that air contained 10 per cent water vapour; he could not have known that particles of air were solid. These were unknown parameters. Though Westfall[81] bluntly states that making these assumptions was 'nothing short of deliberate fraud', my opinion is that making mathematical approximations in an intractable problem is actually the best that can be done to show that a theory is feasible at all.

Nevertheless, other questions remain. When one examines the three editions of Newton's *Principia* (published in 1867, 1713 and 1726, respectively) one notices that the corrections Newton made in his calculations were made *a posteriori*, that is, he knew what the result should have been, and then he adjusted his data until they fitted his predictions. So for instance, in his calculation of the velocity of sound he corrected the air:water density from 1:850 in the first edition of his book to 1:870 in the second. Similarly, Newton's calculation of the distance from Earth to the moon were based on the known values for the gravity constant g. Newton first used the data of Copernicus, Vendelin and Tycho for the distance of the moon, being 60.5 radii of earth. From this the value of $\frac{1}{2}\,g$ came out to be 15 feet and 1 inch, as indeed measured by Huygens. As a matter of fact the distance to the moon was not known, so the choice of the value 60.5 fitted well the g value as determined by Huygens.

The third example of manipulating the data is found in Newton's calculations on Earth's precession of equinoxes, which was known to be about 50 seconds. Westfall[81] had this to say about them:

> In the case of precession . . . the correction of a faulty lemma in edition one, imposed the necessity of adjustment of more than 50 per cent in the remaining numbers. Without even pretending that he had new data, Newton brazenly manipulated the old figures on precession so that he not only covered the apparent discrepancy but carried the demonstration to a higher plane of accuracy. (p. 755)

Newton's manuscript for the second edition of *Principia* was edited by Roger Cotes. In his correspondence with Cotes about corrections Newton wrote (February 1719): 'If you can mend the numbers so as to make y^e [archaic 'yours'] precession of the Equinox 50 ″ or 51 ″, it is sufficient.' Cotes made the necessary corrections and wrote to Newton: 'I am very glad to see the whole so perfectly well stated and fairly stated for without regard to the conclusion I think ye distance of 18.5 degrees ought to be taken & is much better than 17.5 or 15.25 & the same may be said of ye other changes in ye principles from which the conclusion is inferr'd.' Thus in the second edition of Newton's book the results on precession become accurate to within 1 part in 3000.

But surely there is nothing wrong in trying to bring theory and experiment into agreement, provided that the hypothesis is further tested by experimentation, and not presented as the last word. The process of reasoning backwards from an experimental result to correct a detail in a theoretical model need not be considered dishonest.

The following story related by A. H. Boultree[82] illustrates a similar situation this century. Boultree quotes J. C. McLennan, who said during a lecture at the University of Toronto that he had once remarked to Nils Bohr how wonderful it was that his (Bohr's) equation yielded very accurate values of Rydberg's constant (a physical constant related to atomic spectra). Bohr replied: 'Of course, McLennan, I made it come out this way'.

An interesting footnote concerns Newton's jealousy of his work and of priority of his findings.[83] Newton wrote on 16 April, 1676 a letter to the secretary of the Royal Society, warning him that Boyle's invention published in the *Philosophical Transaction* of Royal Society in February of that year could do great social harm. What was this harmful invention? Boyle had noted that heat was produced when mercury was mixed with gold dust. Newton warned the Royal Society of unnamed dangers to the society if Boyle's reaction became known to the uninitiated. It is known at that time Newton himself had been carrying out alchemical experiments, and one may therefore suspect that Newton was afraid Boyle might 'steal' his own priority in discovering the Philosopher's Stone to transmute metals, and therefore used his authority to stop the dissemination of Boyle's findings.[83]

MENDEL - WHO COUNTED THE PEAS?

It took some 40 years for Gregory Mendel, the monk who, in 1865, published a paper on inheritance in the garden pea to become recognized as the father of modern genetics. His revolutionary paper was published in Brno in Bohemia (now Czechoslovakia) and was available at that time at the Linnean Society and the Royal Society in London. Nevertheless, it was largely ignored and rediscovered only in 1900.[84]

The scientists who unearthed Mendel's paper were two European botanists, students of heredity – Corgens in Germany and Tchermak in Austria. They understood that Mendel's discovery was applicable not only to plants, but also to animals and man. The essence of Mendel's discovery was that the units of heredity, later named genes, were transmitted from generation to generation unchanged, and that various combinations of these units were being reshuffled in each generation. There is no question that Mendel's contribution is one of the outstanding advances in the history of biology.

How well did experimental findings support his theory of independent inheritance of traits? Sir Ronald A. Fisher, known worldwide as an authority in biological statistics, examined Mendel's writings in detail.[85] He expressed the opinion that Mendel's talent lay in recognizing the purely mathematical and combinative properties of sets of genetic properties as expressed by two or three factors inherited independently of each other. The independence Mendel was concerned to demonstrate, according to Fisher, was 'closer to a logical than a statistical independence'. It seems that at the time of rediscovery of Mendel's writings the European botanists did not pay much attention to the actual numbers collected in Mendel's experiment, but only to the principles which Mendel derived from these numbers.

In 1936 Sir Ronald Fisher undertook a reconstruction of Mendel's data and statements as to find whether his postulates were indeed plausible. Fisher came to an amazing conclusion: though Mendel's report was to be taken literally, and his experiments could be reproduced just as he described them, some of the figures were inexplicable. The observed results were simply too good! In one of the series of experiments (a second generation of peas that was bred from hybrids or crosses) Mendel tabulated properties such as round or wrinkled seeds and yellow or green colour of their endosperm. Where Mendel expected the ratios of 2:1, the actual experimentally found ratios came out as 1.93:1 and 2.1:1. According to the laws of statistics, this agreement between the predicted ratios and those found experimentally was too good to be true. In another experiment Mendel counted 600 pea progenies. In this case the expected ratio between their traits was 3:1, that is, he should have expected 200 non-segregating plants. The actual counted number was 201, again, too close to the predicted to obey the laws of chance. Fisher therefore concluded:

> An examination of the general level of agreement between Mendel's expectations and his reported results show that it is closer than would be expected in the best of several thousand repetitions. The data have . . . evidently been sophisticated systematically and I have no doubt that Mendel was deceived by a gardening assistant, who knew too well what his principal expected from each trial made . . . This possibility is supported by independent evidence that the data of most, if not all, of the experiments have been falsified so as to agree closely with Mendel's expectations.[85]

Fisher's conclusion indicates that prior to his reported experiments, Mendel was well aware of the independent inheritance of the seven characteristics he studied in peas. Mendel looked upon the numerical frequency ratios as a method of demonstrating the truth of his factorial predictions. The precision with which this system worked made it clear to him what to expect, and how to design an experiment so as to demonstrate the correctness of his hypotheses. The moral Fisher draws from Mendel's contribution to the history of biological thought is that all original papers purporting to establish new facts should be very carefully and meticulously examined.

Fisher's conclusions that Mendel's results were too good to be true were based on a known statistical test of the correlation between the observed frequency of traits and the theoretically expected values. Fisher charitably suggested that Joseph Marsh, Mendel's gardener, may have been the culprit; knowing what to expect, he might have counted the samples accordingly. This suggestion was supported by Sturtevant,[86], Orel[87] and Iltis.[88] There were others who insinuated that Mendel did not count *all* of his samples but stopped when he reached numbers indicating the ratio that fitted the theory.[89] Roberts[90] refuted this suggestion by showing that the number of peas Mendel counted from an indicator number of hybrid plants corresponded well with the predicted yield of such plants. Gardner,[91] Zirkle[92] and Campbell[93] interpreted Mendel's results as an outcome of what we defined in chapter 2 as experimenter bias, i.e. that Mendel saw what he wanted to see.

Discounting these last allegations, one is left with Fisher's analysis. The problem with the statistical test (the chi-square test) is that one member of the equation used is the observed trait. These traits, in order to be evaluated numerically, have to fall into discrete groups of individuals and each individual (object) has to be assigned into one such discrete group. In mathematics there exist groups known as 'fuzzy sets'. Fuzzy sets also exist in biology, when the experimenter cannot really decide whether an individual clearly belongs say to set (a) or to set (b).

Pearl[94] tested the effect of fuzzy sets on results of experiments in genetics. In 1911 he performed a crossing experiment between two varieties of maize, one yellow-starchy and the other white-sweet. In the F_1 generation he expected to obtain a distribution of nine yellow-starchy, three yellow-sweet, three white-starchy and one white-sweet. He let 15 trained scientists assign the progenies to one of these groups, each scientist counting some 512 kernels. The categorization of any particular kernel into a group had to be a result of a subjective decision. No two experimenters reported the same numbers in the respective groups. One can therefore argue that in Mendel's experiment no one but Mendel would have reported exactly the same numbers.

When an investigator encounters objects that are difficult to classify, he can either classify them into an 'indeterminate group' and decide later what to do with this group or he can ignore these individuals. In the first case the numbers in the indeterminate group can then be reclassified and assigned into one of the predetermined sets (ideal categories). It is quite clear that either of these procedures would lead to discrepancies between one observer and another.

Root-Bernstein[95] has investigated the problem further by employing undergraduate students (untrained personnel) to count various types of crosses of maize. The second procedure (when the indeterminate individuals are not counted) gave ratios closer to the ideal expected result. When the first procedure was employed (at the end of the counting the indefinites were reassigned into one of the predicted groups), the results were statistically as unacceptable as those of Mendel. Root-Bernstein therefore believes that Mendel's peas represented a fuzzy set of data which required subjective analysis to be fitted into a set of discrete categories: 'There is subjectivity in the process of inventing categories comprehending nature and there is subjectivity in the process of assigning objects to these categories.'

Mendel himself stated: 'Where one is dealing in a general way with degrees of similarity, account must be taken not only of the traits that stand out sharply, but also of those that are often too difficult to put in words' (quoted by Stern and Sherwood).[96] In another paper Mendel wrote 'seeds damaged during their development by insects often vary in colour and shape; with a little practice in sorting, however, mistakes are easy to avoid'.

One cannot easily reconcile the fuzziness of biological reality with discrete statistical ideality. Thus, though Fisher may have been correct in his statistical criticism of Mendel, the results of the experiments by Pearl and Root-Bernstein support the notion that Mendel was a very careful experimenter and observer, and knew how to classify his material. It seems to me that all insinuations about Mendel's possible unethical behaviour should be discounted.

Fisher's suggestion that Mendel's gardening assistant might have been responsible for the over-accurate count of peas because he knew what Mendel hoped for reminds me of other cases of fraud out of respect and sympathy, which we may pause for a moment to look at here.

When I was working on my PhD thesis in the Department of Microbiology at the Harvard Medical School, a story was told to me by a Professor Eaton about Dr George O.Gey, the inventor of a new method for growing animal cells in tubes inserted in a rotating drum, a method that ensured better aeration of cell cultures.[97] Gey had claimed at that time that he could grow animal cells *in vitro* without serum. In the 1940s this was a very daring claim, because until then nobody had succeeded in growing cells taken directly from animal embryos without supplementing the media (containing salts, amino acids and vitamins) with animal serum, which provided some unknown ingredients required for good growth.

Eaton tried to replicate Gey's technique, but was not successful. He therefore invited Gey to come to Harvard to demonstrate his technique there. Gey arrived, accompanied by his faithful technician who had been working with him for many years. When experiments performed by Gey and his technician worked, while others done by Eaton did not, Eaton, knowing Gey's integrity, began to suspect that the technician was doing something that was helping the cultures along. Eaton therefore locked the incubator with the newly set up cultures every evening and had it opened the next day in his presence. As soon as this precaution was taken, no further experiments succeeded. It became apparent that the technician had helped the cultures by stealthily adding serum at night. The reason for his doing so, as explained by Eaton, was that he greatly respected and admired his boss and knew that Gey was enthusiastic when the experiments worked well, and rather despondent when they failed.

He believed that some outside intervention on behalf of the incubated cells was worthwhile to keep Gey happy.

A similar story was related to me by Professor Albert Sabin, the discoverer of the polio vaccine which bears his name. When he was a guest scientist at the Rockefeller Institute in 1935–7, the director of the Institute, Simon Flexner, consulted Sabin about a manuscript submitted for publication in the *Journal of Experimental Medicine* (the organ of the Rockefeller Institute). The article was written by Klaus Jungenbluth, a bacteriologist from Columbia University. In it, Jungenbluth stated that monkeys infected with poliovirus would not develop paralysis if, soon after inoculation of the virus, the animals were treated with a dose of vitamin C.[98,99] Flexner wondered whether such a result was plausible, and asked Sabin, already an expert on polio infection in monkeys, to consult with Jungenbluth about his experiments. Sabin approached Jungenbluth and both agreed to perform a joint experiment on some 40 monkeys, some treated with vitamin C and some left as controls, all, of course, having been first inoculated with a paralytic dose of poliovirus. Sabin insisted that he personally inoculated the monkeys with the virus.

The results clearly demonstrated that vitamin C, administered at, or after poliovirus inoculation, would not prevent the development of paralysis in 39 out of 40 monkeys. Sabin was thus convinced that vitamin C had no miraculous properties and that it was worthless as a therapeutic or preventive drug in poliovirus infection. Sabin published these results in the *Journal of Experimental Medicine*,[100] where, in a glorious understatement, he simply said: 'There is no apparent explanation to the difference between these results and those reported earlier by Jungenbluth.'

Sabin later discovered that Jungenbluth was an innocent victim of his technician's ruse. The technician, who had been working with Jungenbluth for 25 years, knew exactly what his boss was expecting, and he helped him get the results he wanted by injecting an innocuous solution instead of virus into the animals that were about to receive the vitamin C treatment.

In the same category of fraud out of respect is this story, told by Professor Kihara[101] in his lecture on Lysenko (see chapter 5). An amateur chemist in Japan, fictitiously named 'Ito', claimed he could transform any matter into silver by incineration. He reported this discovery to a professor of chemistry who knew Ito well and believed him to be an upright and honest man. When, however, he permitted Ito to carry out the experiments under strict control, he discovered that the silver had been put beforehand into the substance to be incinerated by Ito's faithful servant, who wanted to make his master happy.

Perhaps also Blondlot's discovery of N-rays (chapter 3) that ended in the discredit of that scientist, was 'engineered' by his faithful assistant.[30]

WHY DID KAMMERER COMMIT SUICIDE?

In the early years after the First World War a Viennese biologist, Paul Kammerer, devoted his studies to the Lamarckian theory that characteristics acquired by parents during their lives can be passed on to their offspring.[102-4] An example of such an inheritance would be the acquisition of the skill to fly an aeroplane by the children of a pilot, or the development of strong biceps by the descendants of a blacksmith. In 1909 Kammerer performed experiments with a black *viviparous* salamander and a yellow-spotted *oviparous* salamander, and claimed that he could make each of these acquire the characteristics of the other. When he kept young yellow-spotted salamanders on black background they tended to lose their yellow markings. Their offspring, when held in black surroundings, were mostly black except for a row of yellow spots along the middle of their backs. When these offspring were kept in yellow surroundings, however, the yellow spots on the back fused into a single line. Such striped forms also exist in nature. A crossing of a naturally spotted salamander with a striped one produced offspring according to Mendelian segregation, but crossing of naturally spotted salamanders with experimentally striped ones did not follow the Mendelian rules.

Next, Kammerer transplanted ovaries of naturally spotted salamanders into the reproductive organs of naturally striped ones. After such a transplantation the characteristics of the offspring depended on those of the true mother. Progeny of spotted ovaries, however, resulting from transplantation into artificially striped salamanders, bore the spot or stripe property of the father. These experiments on ovarian transplantation led Kammerer to consider the possibility of inheritance of characteristics acquired through cells other than the germ cells.[102]

Another of Kammerer's experimental animals was the midwife toad, *Alytes obstetricans*, which differs from other toads by mating on land, rather than in the water. In order to be able to hold on to the slippery skin of the females, the males of those toads that mate in water develop thickened and horny swellings on their thumbs and palms. These 'nuptial pads' are pigmented. The midwife toad, which breeds on land, does not possess these nuptial pads during the mating season. Kammerer forced his midwife toads to breed in water by denying them access to dry ground; after several generations he observed that the males developed nuptial, pigmented pads. Kammerer claimed that this new characteristic was then transmitted to their male progeny.

This claim was hotly disputed by Mendelian scientists, and a prolonged controversy ensued, lasting for several years. One of the protagonists was William Bateson, who objected to Kammerer's experiments and intrepretation in several articles and letters in *Nature*.[105-7] Kammerer himself, [103] supported by another scientist, McBride,[108] defended the Kammerer viewpoint. Bateson had no opportunity to examine Kammerer's preserved specimens of the modified *Alytes* for reasons claimed by some to have been the fault of Kammerer, by others of Bateson. Eventually, an American herpetologist, G. K. Noble, arrived in Vienna in 1926 to inspect the preserved specimen of Kammerer's toad. Noble[109] found that the coloration on the thumb (nuptial pad) was due to the presence of Indian ink injected into the area. He published his findings, which were supported by Przibram.[110] Following that exposure, Kammerer admitted the fraud in a letter to the Soviet Academy of Science in Moscow dated 22 September 1926. He stated that he was personally innocent of the falsification, and that he did not know the identity of the person who was responsible.[111] A few weeks after the publication of Noble's findings, Kammerer shot and killed himself.

The obvious question is: did Kammerer himself inject the ink in order to support his Lamarckian claim, or was that done by one of his collaborators?

Arthur Koestler, in his book *The Case of the Midwife Toad*,[112] expressed his belief that the forgery had been perpetrated by an enemy of Kammerer at some time after 1923, when Kammerer demonstrated this specimen in England at a lecture. Koestler thought, though there is no direct evidence for this, that Kammerer had indeed observed genuine nuptial pads in midwife toads that were forced to breed in water.

The Kammerer affair did not end with the exposure of fraud and Kammerer's suicide. In spite of Kammerer's own admission of the falsification and the evidence provided by Noble, Kammerer's claims and findings were used in the Soviet Union to support a Lamarckist ideology during the Lysenko period (chapter 5). At the notorious meeting of the Lenin Academy of Agricultural Science in the summer of 1948, the academician N. G. Belensky (as quoted by Zirkle)[13] described Kammerer's work on salamanders as evidence for the inheritance of acquired characteristics, never mentioning that Kammerer admitted fraud not only with regard to the specimen of *Alytes*, but also with regard to the salamanders. (Kammerer wrote in his letter that there were other objects ((black salamanders)) upon which his results had plainly been 'improved' with Indian ink). Even, as late as 1953 Western scientists like Fothergill[113] and Mason[114] described Kammerer's data on salamanders as proofs of adaptation to environmental changes. (Fothergill's opinion may be disregarded since he was a very odd mycologist, who believed that Eve was Adam's daughter!) Their opinion was probably formed in good faith and it is clear that in accepting Kammerer's discredited experiments as genuine, official doctrine on Soviet biology was influenced with regard to Lamarckian inheritance. Following the publication by J. Segal of a book on Michurin and Lysenko in 1951[115] the reviewers wrote that Segal was right in criticizing geneticists for making no effort to repeat Kammerer's experiments, but also that the author had failed to mention many investigators who had attempted to repeat the experiments of Lysenko and his followers with negative results.

Kammerer was an unconventional personality, with artistic inclinations, and great verbal skill; he was a flamboyant but dedicated worker. (This dedication to the science of reptiles was such that he named his daughter Lacerta, a lizard!) Koestler, in analysing the course of events while Kammerer was battling against the Mendelians, as represented by Bateson, tried to put the blame on Bateson for allegedly refusing to examine Kammerer's specimen when he had the chance, and for using his authority to crush Kammerer's experimental evidence.

Were Kammerer's Lamarckian views really unorthodox? In the scientific literature of that time the belief in inheritance of acquired characteristics was quite widely spread.[116] Koestler[117] himself entertained a sort of modern evolutionary theory which clashed with the new-Darwinian view: 'Neo-Darwinism does indeed carry the nineteenth century brand of materialism to its extreme limits – to the proverbial monkey at the typewriter hitting by pure chance on the proper keys to produce a Shakespeare sonnet'. Let us assume, for the sake of argument, that Kammerer had indeed successfully induced the production of nuptial pads in his male toads. Does this constitute a case of Lamarckian inheritance? It may well be possible that the gene(s) for the production of these pads, which certainly exist in this species, are inactive in the land-breeding midwife toad, but that under specific environmental stimuli they become expressed. Kammerer's experimental procedure, as described by him, was to take many hundreds of eggs from female toads and to keep them in an aqueous environment. Only a few per cent survived. He then repeated the same procedure for several generations, losing, of course, the majority of individuals in the process. He was therefore applying a strict selective pressure on the developing eggs so as to select those genetic factors that would permit the eggs to develop into adult animals in water. It is thus possible that one of the traits so selected, by keeping the toads constantly in water, was the development of a property particularly adapted to an aqueous environment – namely the production of nuptial pads (Waddington's 'genetic assimilation').

In order to be able to state that the Lamarckian rule operated in the case of the midwife toad, *all* the eggs should have survived in water in the first, as well as in the following generations, and the nuptial pads should have evolved under these conditions and remained hereditarily fixed.

It is indeed unfortunate that an inconclusive experiment, which in any case would not have proved the point Kammerer was trying to make, should have been rigged, and should have led to Kammerer's tragic death and the discrediting of his integrity as a scientist.

CARREL'S IMMORTAL CELLS

At the beginning of this century Alexis Carrel, a surgeon and biologist at the Rockefeller Institute in New York, greatly intrigued biologists by developing a technique to grow animal cells outside the body. He grew fibroblasts taken from the hearts of chicken embryos in flasks, and was able to do so continuously for 34 years. Carrel believed, and led others to believe, that these cells were 'immortal'. Careful investigation by Witkowski[118] into the events surrounding the cultivation of these cells, however, suggested that there was some manipulation of experiments involved.[119]

Alexis Carrel was a pioneer of tissue culture, a method by which pieces of organ and tissue removed from a developing animal can be cultivated outside the body in vessels containing suitable nutritional media. His techniques, based on his expertise as a surgeon, were so complicated as to appear almost magical, certainly to the layman!

Carrel was born in France and obtained his degree as Doctor of Medicine in Lyons. In 1904 he emigrated to Chicago, and from there was appointed to the Rockefeller Institute in 1906 where he worked until 1938. He was one of the few Americans to win a Nobel prize in 1912, for his surgical achievements in joining severed blood vessels.

In the Rockefeller Institute Carrel was responsible for the development of tissue culture techniques together with Montrose Burrows and Albert Ebeling. Other pioneers in this field, Willmer and Paul, criticized Carrel's surgical techniques, which involved wearing black gowns, masks and hoods (as was the current practice in operating theatres at that time to avoid contamination with bacteria, fungi, viruses, etc.) These complicated precautions dissuaded many biologists from following the methods developed in Carrel's laboratory.

Carrel's techniques involved taking small fragments of embryonic chicken heart, placing them on a coverslip in a drop of plasma diluted in water, letting the plasma clot on the coverslip, inverting the coverslip over a hollow-ground slide, and incubating the culture at 39 °C. When subculturing was required, pieces of the tissue culture were removed with a cataract knife, and transferred to a new drop of hypotonic plasma. I still remember how fascinated I was as a high school student reading about Carrel's experiments and his observation in a microscope of beating heart muscles of chick embryos in his cultures. By 17 January 1912, Carrel had established 16 cultures of such embryonic heart fragments. In the next two months 11 of them died, and of the remaining five only one, No. 726, apparently survived until September 1912. The survival of chick cells in culture, outside the body, for several months was then described by Carrel in the *Journal of Experimental Medicine* under the title 'On the permanent life of tissue outside of the organisms'[120] Carrel declared that he had determined 'the conditions under which the active life of a tissue of the organism could be prolonged indefinitely'.

Carrel's collaborator, Ebeling, claimed to have cultivated one such heart embryonic tissue culture at the Rockefeller Institute until 1938. Ebeling later moved to Lederle Co. and took the culture with him; he kept it there until 1946, when it was discarded. Carrel and his co-workers published a number of papers about the culture, one after it had been growing for 16 months, the next after 28 months. By 1919 the 'immortal' tissue had been cultured for 1390 passages. The final paper by Ebeling appeared in 1922 after 1860 transfers.[121] The 'immortalization' of a living tissue attracted public attention and many articles about it appeared in the daily newspapers and in popular scientific magazines, such as *Scientific American*.

Carrel's pioneering efforts in tissue culture had a great impact on the developing field. Albert Fischer, however, who worked in Carrel's laboratory between 1920 and 1927, wrote in his book on tissue culture, published in 1925, that the majority of biologists, morphologists and pathologists did not have much success in trying to repeat Carrel's experiments. 'Consequently, many investigators became sceptic and pessimistic in regard to the employment of the method'.[122] Nevertheless, Carrel's prominence in the field was widely recognized. At the Tenth International Zoological Congress in Budapest (1927), the president of the congress, Professor van Lenhossek, spoke of Carrel as the 'genius' who had developed tissue culture methods. As a leading figure in the field of tissue culture research, Carrel received enormous publicity, especially for the development of the 'immortal cell' strain. It is ironic that Carrel's success in tissue culture deterred many others from following his example because of the difficulties of the method.

Nobody doubted Carrel's results until 1956, when Hayflick demonstrated that embryonic human cells of the connective tissue, the fibroblasts, which kept their normal and constant (diploid) set of chromosomes and did not undergo transformation into cancerous cells, could not be cultivated in culture for more than 50 ± 10 doublings.[123,124] In contrast, cancer cells (like HeLa) have acquired immortality and can be grown outside the body almost indefinitely. Hayflick and Moorehead[125] showed that this limitation of growth was not due to inhibitory substances being released from ageing cells or to the effects of activated latent viruses. Many theories have been proposed to explain the biochemical and biological basis of age-related decay of cells, but the fact remains that in all cases where proper precautions were taken in the cultivation process of diploid cells, their life span, that is the number of times the cells are able to divide, has been limited.

Hayflick's work, as well as the experiments that followed, focused the attention of biologists on the problem of how to explain the apparent immortality of Carrel's cells.[126] By that time Carrel was dead (he died in France in 1944, in disgrace as a Nazi collaborator). It was, therefore, not possible to find out the details from him.

The statement that nobody doubted Carrel's results until 1956 should be qualified. One person who took special interest in Carrel's work in the early 1930s was Ralph Buchsbaum, who became Professor of Zoology in Pittsburgh in 1960 (he retired from the University of Chicago in 1972). Buchsbaum visited the Rockefeller Institute in 1930, at a time when Carrel was away on a vacation in Spain.

Dr R. C. Parker, Carrel's chief assistant was too busy to see Dr Buchsbaum, who spoke, therefore, to Ebeling. He showed him around the laboratories, but would not show him the 'immortal' cells because of the risk of contamination. Many years later Buchsbaum wrote to Witkowski describing his visit to Carrel's laboratory in 1930:

> I could not return to Chicago without seeing the famous immortal strain, so I returned to the floor where I met a young woman technician. I pleaded with her to let me see the cultures. She said Dr Carrel and Dr Parker would have a fit if they knew, but 'what harm could it do to see them?' When I looked at the cells and said they were full of fat globules and obviously on the way out, she said slyly: 'Well, Dr Carrel would be so upset if we lost the strain, we just add a few embryo cells now and then . . . We make new strains for new experiments. Dr Parker says he will retire the strain soon, it costs too much to keep it going.'[119]

Witkowski also had a letter from Dr Margaret Murray (who worked at that time in Carrel's laboratory) in which she related that one of Carrel's technicians was an anti-fascist and very much disliked Carrel's political and social beliefs. This assistant might have been the one who tried to discredit Carrel as a scientist. Other scientists who worked with the immortal cells in the period between 1930 and 1939 were not willing to confide in Witkowski. Thus we are left with Buchsbaum's account only.

Today, the well-established fact that normal, diploid cells are not immortal leaves us with the following hypothesis: Carrel, committed to the concept of the immortality of his cells, repeatedly stated that his cells could be grown indefinitely. This was part of his mystical view of the nature of life; his belief in the unending life force. His belief in the immortality of cells must have been shared by his colleagues in the laboratory, but eventually, when difficulties arose, as described by Buchsbaum, occasional replenishment of the cultures (perhaps even inadvertently by the use of embryo extracts that might have contained some living cells), would have contributed to the concept of immortality. One should not detract from Carrel's achievement to have cultivated chicken embryonic cells for prolonged periods of time in an era when antibiotics were not available to overcome any contaminating bacteria and anybody less careful and meticulous than Carrel or his close co-workers would have quickly lost the cultures by bacterial or fungal contamination.

The immortality of Carrel's cells, however, instead of being a fact, is now only a legend.

THE BURT CONTROVERSY - INHERITANCE OF INTELLIGENCE

Sir Cyril Burt, who died in 1971 at the age of 88, was an eminent British psychologist. He was the first psychologist to receive a knighthood and shortly before his death he was awarded the Thorndike Prize by the American Psychological Society. He held a Chair of Psychology at University College, London until his retirement at the age of 68.

In his numerous publications Burt presented an extensive and almost unique assembly of IQ sets to support his hypothesis that intelligence is determined by heredity. Burt collected most of his data in the period 1913–32 when he was a research psychologist in the London educational system. He was a government adviser in the 1930s and 1940s and he was partly responsible for setting up the '11-plus' system of education, in which, at the age of 11, children were tested and assigned to one of three education levels. The influence of Burt's data and theories[127,128] was such that Arthur Jensen of the University of California suggested in the *Harvard Educational Review* in 1969 that a failure of compensatory educational programme for the racial minorities in the USA might be explained by the hypothesis of dependence of intelligence on racial heredity. This idea also gained other support.[129]

Burt first reported IQ tests done on 21 pairs of twins in 1955.[127] By 1958 he had published data on 30 pairs of twins[128] and in his final paper in 1966 the tests comprised a total of 53 pairs.[130] In these papers Burt found a very high correlation (0.944) between the IQ scores of identical twins reared together, while in the case of twins who had been separated and had grown up in different environments the correlation coefficient was 0.771. (Correlation coefficient is a number expressing the degree of dependence between two variables, in this case between the IQs of parents and progeny. This number may range between 0 and 1, where 1 indicates a perfect correlation and 0 means no correlation at all.) In his final paper on the subject,[130] published when he was already 83, Burt indicated that the foster homes for the separated twins studied were chosen at random and rejected the claim of environmentalists that the high correlation for separated twins had been due to the way foster parents were chosen.

There were scientists who doubted Burt's data and conclusions. Among them was Sandra Scarr-Salapatek of the University of California. She thought Burt's data looked 'funny' and wrote asking for clarification about his procedures. There was no satisfactory answer. Nevertheless, as late as 1971, Dr Richard Herrnstein of Harvard put forward the theory that social standing was based on inherited intelligence and he supported his contention with Burt's data.[131]

In 1972 Burt's papers were brought to the attention of Leon Kamin, a psychologist at Princeton University, who at that time was mainly studying conditioned reflexes. Kamin immediately noticed that Burt's papers contained internal contradictions and lacked methodological data such as the sex of the children tested, the types of tests administered etc.[132] Kamin also noted that in his paper published in 1939,[133] Burt stated that the methods he used 'were described more fully in degree theses of the investigators mentioned in the text' or that they were, according to Burt, buried in inaccessible theses. In another paper, published in 1943,[134] Burt stated: 'A fuller account of sources, of calculations, with detailed tables will be found in her [J. Maver's] degree essay (filed in the Pyschological Laboratory, University College, London)'. In fact, as Kamin discovered, no such essay or thesis was filed or even submitted to the university. In that same paper, Burt wrote also that some of the inquiries were published in London local authority reports or existed as typed memoranda.

At the beginning of his scientific career Burt had published a paper, jointly with Moore, in which he had stated that the more important of his tests were carried out on more than *a thousand* children aged 6–14 (the Liverpool study). In 1939, however, Burt wrote: 'The value and reliability of group testing . . . were demonstrated by Moore, Davies and myself . . . These were, we believe, the first investigations in this country in which the number of children tested ran into well over a *hundred*' [my italics].[133] When Burt was challenged in 1954 in a letter to the *British Journal of Statistical Psychology* about the availability of detailed tables relating to the Liverpool research,[135] Burt answered that 'Mr Moore will himself publish a fuller account of his analysis in the forthcoming issue of this journal'[136] In fact, Moore never published anything after 1919.

The most important of Kamin's findings, in spite of his being an outsider in the IQ testing field, was that in all three of Burt's papers on twins, the correlation coefficients for different numbers of twins were exactly the same, namely 0.771 for separated twins and 0.994 for twins reared together. Amazingly, no one before Kamin had noticed this extraordinary coincidence in numbers. Kamin came to the conclusion that Burt had 'cooked' his data in order to arrive at the conclusion he wanted. Kamin's view was supported by Liam Hudson, at that time Professor of Psychology at Edinburgh University.

On the other hand, Professor Jensen, who had been a post-doctoral student of Eysenck, who in turn was Burt's pupil, did not hold such an extreme view. Jensen greatly admired Burt and considered him one of the world's great psychologists. When accusations about Burt's data were voiced in 1972, after Burt's death, Jensen visited England to collect a complete set of Burt's reprints and to write a review of his work. To his surprise, he found in Burt's papers some 20 instances of invariant correlations, although the sample size of experimental subjects changed, a finding that was almost impossible to explain as occurring by chance alone. In 1974 Jensen wrote: 'It is almost as if Burt regarded the actual data as merely an incidental backdrop for the illustration of the theoretical issues in quantitative genetics, which, to him, seemed always to hold the centre of the stage'.[137] Jensen's mentor, Eysenck, also agreed that Burt's data were unusable and unreliable, though he did not go so far as to accuse Burt of concocting them. In his obituary note on Burt, Eysenck expressed his surprise at the revelations since he knew that Burt was 'a deadly critic of other people's work when this departed in any way from the highest standards of accuracy and logical consistency.[138]

We thus face the question of whether the 'errors' in Burt's papers were due to inattention to detail on the part of a 72-year-old scientist (the view of Jensen and Eysenck), or to a deliberate attempt to deceive (Kamin). Kamin went so far as to suggest that Burt invented all his data, from the very beginning of his career in 1909.[139] Jensen, on the other hand, doubted the feasibility of this statement; he thought that if Burt had been trying to fake the data, as a person with high statistical skills, he would have done a better job.

The controversy persisted after Burt's death and seemed to demand further investigations. It was thought that the inspection of Burt's original notebooks would provide the answer, but unfortunately such notebooks and raw data were not available. After Burt's death some of his colleagues took books and reprints from his house, but left behind some six tea chests filled with test sheets and notebooks. Burt's housekeeper, acting upon the advice of some of his colleagues who had been asked what to do with this material, burned the chests with all their contents. In such a situation raw facts cannot be established and the judgement of Burt's integrity as a scientist has to depend on conjectures based on his published work.

That Burt had been inventing data appeared more probable after it was discovered that he had published his critical papers with two collaborators, Miss Margaret Howard[140,141] and Miss J. Conway.[142] These two collaborators could not be traced, and in October 1976 Oliver Gillie of *The Sunday Times* newspaper, who researched the matter thoroughly, reported that the two women did not exist. A possibility was entertained that in the early 1930s at least Miss Howard was in Burt's entourage, but it was quite certain that the women could not have collaborated with Burt in his crucial papers published in the late 1950s.

The names of Howard and Conway also appeared in the pages of the *Journal of Statistical Psychology* (edited by Burt) as authors of book reviews praising Burt's publications, indicating Burt's priority in many discoveries and criticizing the publications of his opponents. After Burt had ceased to be the editor of the journal, no additional reviews by Howard and Conway were published. Those who knew Burt well felt that the reviews were written by Burt himself, in his unmistakable style, and that he used the names of Howard and Conway as pseudonyms.

A further suspicious circumstance was that after 1950 Burt, having retired from the University, could not have administered the IQ tests himself and must have had collaborators for testing the twins, especially when travel was needed to the places where the twins lived. When asked about this in 1969 Burt wrote to a correspondent that he delegated the job to Conway and Howard.

Amazingly, during Burt's lifetime his work was never challenged, despite its shortcomings. He was a very powerful figure at the time of his retirement, and remained influential after that. More importantly, Burt's results for IQ correlations were not only in accordance with other studies (Figure 5) but fulfilled the general expectations and beliefs about inheritance of intelligence prevalent in the psychological community.

So, the question remains whether, as Kamin believes, Burt's data were fraudulent, or whether as Eysenck suggests, that Burt, now old and ill, merely carried over the correlation figures from his earlier papers in order to avoid the difficult task of recalculating the new data. Such a procedure would not constitute fraud, since it would have been done without the intent to deceive. Nicholas Wade, who reviewed Burt's story in *Science*,[143] finds it hard to believe that the combination of implausibility in Burt's results, the apparent use of pseudonyms and the failure to locate Misses Howard and Conway speak for Burt's innocence: he seems to come down on Kamin's side.

In 1978 came a very scholarly analysis of Burt's work by Dorfman of the Univeristy of Iowa.[144] He analysed in detail Burt's paper of 1961[145] on intelligence and social mobility, and concluded that Burt's data 'were fabricated from a theoretical normal curve, from a genetic regression equation and from figures published more than 30 years before Burt completed his surveys'. Dorfman also showed that Burt, instead of providing new data, copied the figures from tables

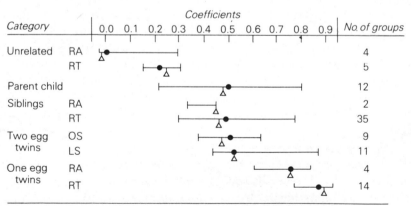

Full circles — median correlation coefficient; open triangles — Burt's data
RA = reared apart; RT = reared together; OS = opposite sex; LS = like sex

Figure 5. Comparison of Burt's data on inheritance of intelligence with data from similar studies on twins conducted by others. Median values indicated by triangles (based on Erlenmeyer-Kimling, L. and Jarvik, L. F. 1963: *Science*, 142: 1478).

published in his paper on vocational guidance in 1926,[146] which in turn were based on figures from a census taken in 1921.

Even if serious doubts are entertained about the validity of Burt's data, the fact remains that the deletion of these data from the overall picture of inheritance of intelligence, based on some 52 other studies, would have no appreciable effect. These 52 studies, reviewed by Erlenmeyer-Kimling and Jarvik,[147] comprised 1082 monozygotic and 2052 dizygotic twins, 256 siblings reared apart and 15,513 siblings reared together, 7883 parent–child pairs and 482 unrelated pairs. The authors concluded that there was a 'marked trend toward an increasing type of intellectual resemblance in direct proportion to an increasing degree of genetic relationship regardless of environmental communality'. The overall correlation coefficient for twins reared together was 0.87 and for those reared apart, 0.75. Burt's figures differ from the median (a statistical average) values obtained by other authors in an unsystematic way; Burt's corresponding median values were 0.94 and 0.77 respectively.

One can therefore repeat the statement that Burt, having once calculated the correlation coefficients for 'intelligence scores', and being familiar with the literature, seems to have published his data repeating the earlier calculated correlation. Burt's apparent forgery was detected because the variation in his results was below the likely vagaries of chance.[118] Could Burt's actions be classified as fraud? According to Edwards,[149] 'fraud occurs when the conclusions of a scientific research are blatantly at variance with the known facts; when lying is plausible, it is a misdemeanour'. Burt's action would thus be classified as misdemeanour. This is a view I shall dispute in chapter 14.

What is quite astonishing is that although the figures reported by Burt (0.944 for twins reared together, 0.77 for those reared apart) were the same for groups of 21, 'over 30' and 53 pairs of twins, figures that were recognized by Kamin as being incompatible with the known statistical rules, it took the British Psychological Society a further six years to reach this same conclusion. Estling writing in 1982[148] concludes: 'Burt's crime is the very plausibility of his fiction, which was manufactured to feed his, and our prejudices . . . for heritability.'

HOW MILLIKAN WON THE NOBEL PRIZE

Robert A. Millikan, a famous American physicist of the University of Chicago, won the Nobel prize in 1923 for his experiments of 1910 in which he determined the electrical charge of the smallest component of the atom, the electron. In his Nobel lecture Millikan said:

> My own work has been that of the mere experimentalist whose main motive has been to devise, if possible, certain crucial experiments for testing the validity or invalidity of conceptions advanced by others regarding the unitary nature of electricity . . . The success of the experiments first performed in 1909 was wholly due to the design of the apparatus, i.e. to the relation of the parts . . . scarcely any other combination of dimensions, field strengths and material could have yielded the results obtained.

What were these experiments? They essentially followed those of Regener carried out in Thompson's laboratory in Cambridge (England). Water droplets were produced in an expansion chamber between electrically charged horizontal plates; the retardation of the rate of fall of these droplets, as determined by microscopical observation, was related by Stoke's law (describing the fall of very small particles) to the electrical charge on the droplets, which they acquired in the electric field between the plates. Initial results, which Millikan reported at a meeting in Canada in 1909, gave values within an exact mutiple of the smallest charge on a droplet (this charge was calculated from the known strength of the electric field created by the condenser plates); this smallest electrical charge was calculated to be 4.65×10^{10} electrostatic units. The difficulty with experiments was that the water droplets evaporated so quickly that observation of their fall was possible for a few seconds only.

At this stage of research, a graduate student, Harvey Fletcher, joined Millikan's group. Following some discussions in the group on the difficulties encountered with the water droplets, Fletcher suggested the use of oil droplets, which would not evaporate. Within a day Fletcher had set up an apparatus consisting of an arc lamp, an atomizer for spraying oil, two brass plates (the upper one with a hole in the centre to permit the entry of the falling oil droplets), a telescope and batteries to charge the plates. Fletcher wrote in his posthumously published memoirs

> I then tried out the apparatus. I turned on the light, focused the telescope, sprayed oil over the top of the plate and then came back to look through the telescope. I saw a most beautiful sight. The field was full of little starlets, having all colors of the rainbow . . They executed the most fascinating dance, I had never seen Brownian movement before. Here was a spectacular view of it.[150]

Fletcher then applied an electrical field to produce 1000 volts across the plates and looked again through the telescope. He noticed that some droplets fell and others moved upwards. 'I knew some were charged positively and some negatively . . . I spent the rest of the day playing with these droplets and got a fairly reasonable value of *e* before the day ended.'

Millikan, who was away from the laboratory that day, only saw Fletcher's set-up the next day, and was very much excited. From then on he continued working closely with Fletcher. They made some improvements in the design of the apparatus and within six weeks the discovery was announced publicly. Fletcher thought the findings would be published in a joint paper with Millikan. Millikan, however, explained that if Fletcher wanted to use a published paper for his doctoral thesis, he must be its sole author (such were the regulations of the universities in those days), and since the oil-drop experiments were a collaborative work with Millikan, there was a problem. The problem was solved by the friendly advice (or was it demand?) of Millikan, and so the paper, entitled 'The isolation of an ion, a precision measurement of its charge and correction of Stokes' law, was published in 1910 by Millikan alone.[151] In the text of this paper, however, Millikan gave credit to Fletcher:

> . . . Mr Harvey Fletcher and myself, who have worked on these experiments since December 1909 and have studied in this way between December and May from one to two hundred drops which had the initial charge from 1 to 150 and made from oil, mercury and glycerine, and found in every case the charge on the drop to be an exact multiple of the smallest charge which we found that the drop caught from the air'.

On the issue of authorship of the paper, Fletcher wrote in his memoirs:

> I did not like this but I could see no other way out, so I agreed to use the fifth paper as my thesis . . . It is obvious that I was disappointed as I had done considerable work on it and had expected to be a joint author. But Millikan was very good to me while I was in Chicago . . . and through his influence I got into the graduate school.

Fletcher insisted that he bore no ill will toward Millikan for not letting him be a joint author on this first paper, which effectively led to Millikan's being awarded the Nobel prize.

During the period 1911–13, Millikan had a foreign rival, Dr Felix Ehrenhaft of the University of Vienna, who also measured the charges on oil droplets. Ehrenhaft's experiments, however, indicated that there were subelectronic fractional charges on the droplets, rather than the exact multiples of unit charges postulated by Millikan. How was it that the two laboratories obtained conflicting results?

Holton[152] and Franklin[153], who in 1978 and 1981 studied Millikan's unpublished data on his oil drop experiments during that period, found that Millikan graded his results in his notebook from 'best' to 'fair'. A paper published by Millikan in 1913 was based, according to the notebook, on 140 observations, but he had excluded 49. Nevertheless, Millikan stated in the published paper that the results 'represented *all* the drops experimented upon . . .' It seems therefore that, contrary to his statements in the paper, Millikan was rather selective in the use of his experimental data. Thus, in distinction from the data of Ehrenhaft, which contained a wide range of fractional values, Millikan's data came out in his 1913 paper as very elegant and clear. It was this paper that put an end to the controversy with Ehrenhaft, who ended his life in disillusionment while Millikan had the recognition afforded by the Nobel prize.

It is ironic that in a paper published in *Science* in 1923, Millikan wrote '4/5 of all experiments which we make in our physical laboratories in the hope of developing new relations, establishing new laws, or opening up new avenues of progress, are found to be directed along wrong lines and have to be abandoned.'[154]

There are two issues involved in this story. I should like to separate the conduct of Millikan as the thesis supervisor of Fletcher, from his later act of omission of some results in his published papers.

The mentor–student relationship is a very complicated one. The attribution of credit for a new idea or discovery made by a professor and student, or by the student under guidance of his supervisor, or by the student entirely on his own, cannot be viewed as a black and white situation.

Let us recall the famous Waksman–Schatz discovery of streptomycin. At that time the only known and clinically used antibiotic was penicillin, which had been discovered some 14 years previously. Selman Waksman, at that time Professor of Microbiology at Rutgers University in New Brunswick, New Jersey, had been studying a class of bacteria-like ubiquitous soil fungi, the Actinomycetes. Knowing that penicillin was produced by a mould, Penicillium, Waksman assumed that other classes of moulds or fungi might also be producing antibiotic substances that would enhance their survival in soil. He therefore set up his students to isolate as many Actinomycetes as possible from soil and screen them for antibiotics against disease-causing bacteria. One of his doctoral students, Albert Schatz, made the momentous discovery that a fungus called Streptomyces actually produced an antibiotic that would kill bacteria causing tuberculosis. This antibiotic also proved effective against other human and animal diseases. The paper describing the discovery of streptomycin was signed by both Waksman and Schatz. Nevertheless, Waksman alone was awarded the Nobel prize for this discovery.

When Waksman patented streptomycin and obtained substantial royalties from the manufacturers (which incidentally were mostly used to build and finance the Waksman Institute of Microbiology), Schatz sued Waksman and demanded a share of royalties from the sale of streptomycin. The matter was eventually settled out of court, but the precedent that a student may sue his professor incensed the scientific community to the extent that Schatz was ostracized in the USA and could not obtain a research position there. (He went to South America and became engaged in education.) The Nobel Prize committee recognized the fact that behind Schatz's discovery was a general concept based on the arduous studies of Waksman, and that this background, which Waksman provided, was essential for Schatz to make the discovery. The Nobel Prize was, however, given to Waksman only.

In the Millikan–Fletcher case there was, after all, an amicable settlement between the two: Fletcher, though feeling wronged, agreed out of his feeling of respect and gratitude to Millikan, to let his mentor to reap the glory. Here was a situation where a young student joined a team that was set up to attack a certain physical problem. The basic idea to measure the charge of the electron on droplets was already there when Fletcher arrived. True, the idea to use oil rather than water was a real breakthrough and one would expect that Fletcher be given credit for this idea and appear as a co-author in the article. Unfortunately, the strict rules applying at that time to PhD students (in some universities they still apply now), that the thesis has to be the student's own work, caused difficulties. Thus a situation developed in which Millikan, as the supervisor, reaped the harvest alone.

I should like to think that the problem would be solved differently today, although I personally know of cases where graduate students are not permitted to pursue their own ideas and have to toe the line set for them by their supervisor. There are still many laboratories (mainly in the biological and medical sciences) where the rule applies that any paper coming from a student should be signed by the head of the laboratory (usually as the last author). I have hardly encountered any comments about this procedure that would classify it as unethical. On this basis, then, Millikan's behaviour toward Fletcher would not be considered unethical; nevertheless, I believe that taking away the credit from a student who has had an innovative idea and executed it should be censured.

What of the other charge against Millikan, that he left out some of the experimental data assembled in the course of measuring the charges on the electrons? Is it improper to leave out data? In the case described, Millikan left out one-third of his measurements from the published report. Salvador A. Luria, also a Nobel prize winner, expressed his conviction that 'leaving out data that inexplicably conflict with the rest of scientists' data or with the proposed interpretation is an anathema'.[16] From this point of view, which I share, Millikan's selective reporting of his results would not conform with the norms and ethics of science, and would be branded as misconduct.

5

Lysenko – Science and Politics

Trofim D. Lysenko, a Soviet agronomist active during the period between 1929 and 1965, won the highest Communist Party support for an effort which, in essence, was based on the annihilation of the science of genetics. The Marxist ideology was founded on the concept of the malleability of human nature. In support of this theory Lysenko applied his energy and drive to show that the nature of plants can also be moulded by the environmental conditions, notwithstanding their genetic character.

Lysenko rebelled totally against modern-day science. His multifaceted activities involving the top levels of Soviet agriculture stemmed from his arrogant belief that he knew better than the academic scientists how to increase the yields of agricultural products. This led to a situation where Soviet agriculture was abused for 35 years.

Through him many leading scientists lost their jobs, and sometimes their lives, after having been accused of 'wreckage of Soviet agriculture'.

No true scientist, Lysenko never addressed any genuine scientific problem which he and his followers considered a scholastic obstacle to quick solutions in agriculture. Lysenko attended a secondary school of gardening (a type of horticultural college) and subsequently worked at a rural experimental station belonging to the Kiev Agricultural Institute. He had no postgraduate training or higher degree.

Lysenko was assigned to a remote agricultural station in the North Caucasus, and there set about trying to find a good winter crop. During that period he re-discovered so-called 'vernalization', a process that involves the moistening and chilling of seeds or seedlings during the winter. When such plants are sown or planted in the spring, they complete their life cycle in a shorter period of time; in regions where the summer is short, such vernalized plants can be harvested before autumn[155]*

N. A. Maksimov, a leading Soviet plant physiologist at that time, said that the results obtained by Comrade Lysenko did not represent anything new in principle: they were not a scientific 'discovery' in the precise sense of the word.[156] The transformation of winter to spring habit by moistening and chilling the seeds was known in the USA in 1857, and was even reported in a Russian agricultural paper in 1885.[157] A German plant physiologist, Klebe, wrote about the phenomenon in a book *Willkuerliche Entwicklugsaenderungen* published at the beginning of this century; this book was then translated into Russian by Timiriazev. Chouard[157] has reported that other studies on vernalization were made in 1918 by Gassner, who did not think the method had any practical advantage.

Lysenko, incensed by the criticism that he had made no 'discovery', and being unable to explain how his vernalization method actually worked, insisted, from 1923 onwards, that *all* varieties of wheat, both winter and spring, would respond to chilling or soaking by hastening the onset of earing, and believed that this would obviously increase the yields. The spring varieties were soaked and kept under controlled temperatures and humidity. They were thus sown in a swollen condition. The procedure was supposed to shorten the vegetative stage. In fact, the exact conditions required for this process could not easily be reproduced on collective farms and in some cases the yields of the vernalized wheat were actually lower.

In 1929 Lysenko was transferred to the Odessa Institute, and within a year, from there to the Moscow Institute of Genetics. From there he reported his work not in established scientific periodicals, but rather in interviews with reporters from mass-circulation newspapers. Lysenko would not submit his articles and data to the scrutiny of journal reviewers.

Lysenko applied vernalization not only to seeds, but also to tubers and cuttings, and claimed success in his methods. He never acknowledged the accumulating data on the influence of plant hormones, claiming that he had disproved these phenomena. (In fact, we now know that hormones control plant growth and the reproduc-

---

* The references in this chapter are taken from the source books *The Rise and Fall of Lysenko* by Zhores Medvedev (1969) and *The Lysenko Affair* by D. Joravsky (1970) and they are presented as they appear in these books.

tive process of plants. There exists today a multi-million pound industry for the production of synthetic hormones for many uses in agriculture, ranging from ripening fruits to inducing rooting of cuttings.) In 1939 Lysenko wrote:

> The speeded up development of such plants (sprouted) we explain basically not by the fact that the eyes of the tubers are sprouted before planting, but by the fact that the sprouts are subjected to the influence of certain external conditions, namely the influence of light (a long spring day) and of a temperature of 15–20°C. Under the influence of these external conditions (and that precisely is vernalization), in the potato tuber's eyes, as they start to grow, there occur those quantitative changes which, after the tubers are planted, will lead the plant to more rapid flowering, and hence to more rapid formation of young tubers.[158]

This represents an explanation without substance. From Lysenko's many publications and pronouncements one would understand that vernalization was the initial stage of development in any plant or its part, and that certain conditions of air, moisture and temperature were essential for the onset of the next stage and for flowering.

As the years went by, Lysenko became more and more powerful and reached the peak of his influence in the Soviet establishment in the years 1948–52, when he was unequivocally supported by Stalin himself. At the 1948 session of the Lenin All Union Academy of Agricultural Sciences (LAAAS). Lysenko made this chilling statement: 'The question is asked in one of the notes handed to me: What is the attitude of the Central Committee of the Party to my report? I answer: ' "The Central Committee of the Party has examined my report and approved it." '[59] Lysenko received a standing ovation for that statement. By that time he had succeeded in destroying the science of genetics in the Soviet Union completely, and had had the most important geneticists removed from their posts; in some instances they had been arrested and even executed.

Consider the fate of Maksimov, for example. Lysenko's paper (with Dolgushin) on vernalization was first presented at the Congress of Genetics, Selection, Plant and Animal Breeding, held in Leningrad in 1929. At that meeting Maksimov also presented a paper on physiological methods of regulating the length of the vegetative period in plants. Maksimov, who at the time was the head of the Physiological Laboratory at the Institute of Applied Botany, had used the method of germination in the cold since 1923; he obtained crops of winter varieties in the first year without damaging the seeds by winter weather. Lysenko instigated a campaign against Maksimov, and 1934 he was banned to Saratov.

In his speeches and lectures Lysenko mixed the 'branch of science' of vernalization with the class struggle, against the so-called scientists and 'kulaks'. He called geneticists and farmers 'class enemies' and thus gained the support of Stalin who, at the Party Congress in 1935, exclaimed: 'Bravo, Comrade Lysenko, bravo!'

The classic concept of heredity as enunciated by Mendel, Morgan, deVries, Goldschmidt and Muller outside the USSR, and by Vavilov and Kolt'sov within the Soviet Union, was upheld by Russian scientists until 1935. Gregor Mendel had shown that genetic traits were inherited as independent units. He saw that the offspring carried a mosaic of such units, which remained separable, and thus foreshadowed the existence of genes. Thomas Hunt Morgan of Columbia University, at the beginning of this century, used the fruit fly Drosophila to demonstrate the 'chromosomal' basis of heredity, i.e., the dependence of heredity on genes located in the nucleus of the cell. He further demonstrated the linear arrangement of genes on chromosomes. The chromosomal concept entailed a fixed number of chromosomes in each species and predicted that changes in the genes in the germ cells (mutations), if viable, would be transmitted to progeny as new traits.

In 1936 Lysenko and Prezent (a lawyer by education, who considered himself a specialist theoretician on Darwinism and on the teaching of natural sciences in secondary schools) rejected the chromosome theory, and announced a new concept of heredity in *Yarovizatsiya* ('vernalization'), a journal edited by Lysenko, as well as in the more popular journal *Sotsrekonstruktsiya Selskogo Khozaistva*. Lysenko and Prezent stated that the idea that chromosomes contain the heredity substance separate from the rest of the organism was an invention of geneticists (called by him Mendelists) and had no basis in experimental facts. One of Lysenko's main arguments in support of his anti-chromosome view was based on his experimental transformation of the winter wheat variety Kooperatorka into a spring form. According to Lysenko's own description[160] this experiment involved a single plant, off-spring of a single individual, a single seed, and was never replicated.

Even if one assumes the correctness of this irreproducible experiment, it still does not contradict the chromosome theory of heredity, because the capacity of plants to change from form A to form B may *per se* be genotypically determined. 'The hereditary base', according to Lysenko, 'is the cell which develops and becomes an organism. In this cell different organelles have different significance but there is not a single bit that is not subject to evolutionary development'.[161] Lysenko called *his* genetics 'Michurinist', named after one Michurin, a practical plant breeder who achieved remarkable success in grafting fruit trees.

At the 1948 meeting of the Lenin Academy of Agricultural Sciences, one of the Lysenkoites, Belensky, said: 'No special hereditary substance exists any more than does the substance of combustion, phlogiston, or the substance of heat, caloric.' The fallacious idea was gaining strength.

Even as late as 1962 similar nonsense was being propagated.

> The hypothetical connection of the empty abstractions of the gene theory with specified substrates–chromosome, DNA–declared to be the natural carriers of heredity does not confer on these abstractions material content, any more than superstitious deification of objects makes the superstitions materialistic.[162]

Strangely, in 1937 the Nobel prize winning geneticist Hermann J. Muller, who had worked in the Soviet Union with Vavilov in the period between 1933 and 1937, and should surely have noticed the Lysenkoite trends, observed 'with satisfaction' that much attention was paid in Russia to the connection between genetics and practical plant and animal breeding.[163] Later events proved him quite mistaken. The acceptance of the views of Lysenko and his followers influenced the Community Party establishment. In 1936, the director of the Institute for Human Heredity, Salomon Levit, was accused of abetting Nazi doctrines; he was arrested and his Institute closed. Levit died in prison. Similarly, Israel Agol, professor of genetics at the Kiev Academy of Science, and the geneticist Max Levin were arrested at that time. Meister, co-director of the All Union Institute of Grain Culture in Saratov, simply disappeared in 1937, and his post as a member of the Academy of Science was given to Lysenko. Among those arrested at that time were Levitsky (an authority in cytology), Govorov (founder of collection of leguminous plants) and Kovalev (a leading fruit breeder).

Of the 35 members of the Institute of Genetics, 31 refused to accept Lysenko's doctrines; most of them lost their positions at the Institute. The chairman of the Department of Genetics at Leningrad University, Karpechenko, was arrested. By 1948 five geneticists had converted to 'Lysenkoism', 22 were repressed* and some 300 had

---

* 'Repressed' meant in the Soviet Union extra-legal punishment for non-existent crimes, or arrest followed either by execution or internment in a concentration camp, or exile to a specific place with or without corrective labour.

been forced into other types of work. The total number of geneticists and non-Lysenkoist biologists 'repressed' reached 77.

The most tragic was the story of the famous Soviet geneticist Vavilov. He courageously continued to denounce Lysenko's pseudoscience. In 1940 he was arrested while on a botanical expedition in the Western Ukraine. Following his arrest, Vavilov was denounced by some of his former colleagues (Yakushin, Vodkov and Shundenko), and was found by the court to be guilty of rightist conspiracy, spying for England, sabotage of agriculture, links with White emigrées etc. and was condemned to death. This sentence was later commuted to a ten years' imprisonment. In 1943 Vavilov died in prison, presumably of malnutrition (though the death certificate mentioned pneumonia). While Vavilov was in prison, the Royal Society in London elected him a foreign member.[164]

But what was the effect of Lysenkoism in practice? How did Russian agriculture fare under his regime? Lysenko's first sensational announcement on vernalization was followed in 1935 by a telegram (!) to the Commissar of Agriculture (Chernov) and to the president of the LAAS (Muralov) claiming the development of a new variety of spring wheat, obtained by artificial cross-fertilization among members of the same variety of wheat. All these new varieties failed the variety testing system, and one, No. 1163, which on Lysenko's insistence was adopted in 1936, was found to be poorer than the standard strains, and was soon forgotten. Nevertheless, during three years of propaganda Lysenko became famous as an innovator of super-rapid production of new varieties. A similar idea – the intravarietal cross-fertilization of rye[165] which was adopted in practice in 1948 – had to be abandoned after a few years because of falling yields and loss by contamination by other varieties.

Declining yields of vegetatively propagated potatoes in Russia was most often due to viral diseases. Lysenko would not accept this notion, however, and claimed that summer planting of potatoes would prevent the decline. (It is true that this practice is of advantage in some southern regions where the rain falls only in autumn, and then helps good tuber formation after the heat has disposed of the virus). After some limited success in the Ukraine in the years 1934 and 1935, large areas went under cultivation by this method with results that soon led agronomists to abandon it.

Another of Lysenko's brain waves was the planting of winter wheat in Siberia on stubble not ploughed under: he claimed he could develop frost resistant varieties for that purpose, but was not successful in this venture. Unfortunately, it seems that among the agricultural specialists were some unprincipled opportunists who reported false data to support Lysenko's contention that the method was successful. Finally, after disaster on hundreds of thousands of hectares, the method was abandoned.

Among other of Lysenko's ideas for the improvement of Soviet agriculture one may mention the growing of sugar beet in summer in Central Asia (1943/44) and the cluster planting of trees. Each time the sugar beet shoots perished. The idea of cluster planting was based on rejection of the Darwinian concept of competition between individuals of the same species (discussed at greater length below) and its purpose was to create protecting belts of forest around fields. Not until 1954 did it become evident that this idea was also a failure, and the All Union Conference of Foresters voted it down as bankrupt, after a loss of several billions of roubles.

Lysenko advocated a new method of soil fertilization using a mixture of chemical fertilizers and manure, and later, of a mixture of 80 per cent soil and 20 per cent fertilizer. Lysenko, ignorant in chemistry, advocated for many years a mixture of superphosphate and lime, a method which is, of course, absurd, because in such a mixture insoluble calcium triphosphate is formed. Since the superphosphate is actually manufactured from calcium triphosphate, sulphuric acid and phosphoric acid the lime treatment of superphosphate changes it back to its starting material – calcium triphosphate and thus the whole effect is wasted. When in 1955 the Technical Council of the Ministry of Agriculture showed, on the basis of hundreds of experiments, that Lysenko's proposals were worthless, Lysenko complained to the Minister of Agriculture, Matskevitch, and the decision of the Technical Council was shelved.

Lysenko's ideas on successes of vegetative hybridization (by grafting) and the production of branched wheat caused much anguish among practical breeders, who had little choice when ordered to obey Lysenko's methods and instructions. Lysenko and his followers claimed that plant hormones did not exist; that they were the invention of idealists: 'The hormonal theory of development, is in fact a mirage, which should be finished off as soon as possible.[166] Because of this attitude to hormones, Lysenko's colleague and close supporter, Prezent, would not accept, on theoretical grounds, the practical method of superovulation in sheep by hormonal stimulation. Zavadovsky, who developed this method, was dismissed, his laboratory shut and the method stopped. Only eight years later, when the Soviet Minister of Agriculture witnessed the success of the method in England, was it reintroduced to the Soviet Union. By that time Zavadovsky was dead.

The success of hybrid corn in the USA, developed by crossing of inbred lines, was derided and ridiculed by Lysenko. At the 1946 meeting of the LAAS, the Lysenkoite Feiginson actually assured the audience (and the Soviet public) that the hybrid corn was a swindle by the Morganists to serve the interest of capitalist seed firms. It took six years before the CEC (Central Executive Comittee) passed at its plenary session a resolution to adopt the American experience.

The final excess of Lysenkoism was the so-called Jersey cattle experiment. Lysenko claimed that crossing of small bulls endowed with the property of producing female offspring that produce milk with high butterfat, with a large cow, would produce offspring in which the father's high butterfat property would dominate. 'We surmise that the zygotes, the embryos from crosses of large cows with bulls of small breeds, will, with abundant nutrition, develop along the lines of the small breed.[167] This belief of Lysenko was based on teleological premises rather than on experimental findings. Nevertheless at the plenary session of the Council of People's Commissars in February 1964 Lysenko demanded that his cow-crossing method be adopted on all Soviet farms, using bull calves from Lysenko's experimental farms. He was supported in this demand by Khrushchev, and the idea fizzled out only because Khrushchev was removed from power later that year.

It can be seen then, how Soviet agriculture suffered from Lysenko's unsound ideas. Unrelenting, also, was his attack on accepted scientific theory. In 1948 he propounded an extravagant idea on the origin of species. New species were not the progeny of parents of the same species, but stemmed from an entirely different, though related, species. In the journal *Agrobiologia* there appeared in the years 1950–55 numerous articles by Lysenko's collaborators claiming transformation of wheat into rye, barley into oats, peas into vetch, vetch into lentils, cabbage into swedes, firs into pines, hazelnuts into hornbeams, alders into birches and sunflowers into strangleweed. All these communications lacked normal, controlled proofs, and were unreliable throughout. Nevertheless, Lysenko stuck to the theory of transformations until 1961.

An interesting reference to the transformation of hard wheat (with a quadruple complement of chromosomes) to hexaploid soft wheat is provided by the famous Japanese wheat researcher and pioneer in the genetics of wheat Hitoshi Kihara.[101] At the 39th Indian Science Congress in New Delhi, Dr. P. Maleshwari of the University of Delhi asked Kihara to review a paper by Lysenko and Karapetian describing an experiment in which a 28-chromosome hard wheat sown late in the autumn produced a few plants that in two or three generations were transformed into a 42-chromosome bread wheat. Lysenko wrote in this paper:

> It has been observed year after year when cultivating branched wheat (*Triticum turgidum*) on experimental plots of the Lenin Academy of Agricultural Science in the USSR and in a number of other localities that admixtures of soft and durum wheat, oats, 2-and 4-rowed barley and also spring rye appear in the crops. All our observations lead us to conclude that the original source of the admixtures was the branched wheat itself. (Quoted by Kihara)[101]

Kihara generously considered that at the root of the 'queer phenomena' observed by Lysenko and Karapetian 'may be an intentional or unintentional misinterpretation of the philosophies of Hegel and Marx. Lysenko may have been a victim of perpetrators, or of a delusion system aided by unscrupulous collaborators; he may have misused science as a political weapon.'

As I have said, Lysenko rejected the Darwinian cornerstone concept of intraspecific competition. In 1956 he came out publicly against this concept and denied its existence among plants and animals in nature. Many scientists in the Soviet Union at that time viewed the controversy in genetics and Darwinism as a genuine scientific debate,[168] but soon it became evident that the position of Lysenko and his followers was far-fetched, based on few facts, and bordering on complete falsification of science. Lysenko's anti-Darwinian arguments (first published in *Literaturnaya Gazeta* in 1947, and then in *Agrobiologia* in 1949) ran as follows: 'There is no intraspecific competition in nature. There is only competition between species: the wolf eats the hare, the hare . . . eats grass. Wheat does not hamper wheat, but couchgrass, goose foot . . . are members of other species and when they appear among wheat . . . they take away the latter's food . . .' In *Agrobiologia* (1949) Lysenko stated:

> Bourgeois biology, by its very essence, because it is bourgeois, neither could nor can make any discoveries that have to be based on the absence of intraspecific competition, a principle it does not recognize. That is why American scientists could not adopt the practice of cluster sowing. They, servants of capitalism, need not struggle with the elements, with nature! They need an invented struggle be-

tween two kinds of wheat belonging to the same species. By means of fabricated intraspecific competition, 'the eternal laws of nature', they are attempting to justify the class struggle and the oppression, by white Americans, of Negroes.[169]

Note here how social and political concepts have now openly replaced scientific considerations.

By 1961 Lysenko was at the height of his political power. When, in that year, the Leningrad University convened a conference on experimental genetics with more than 100 papers to be presented, Lysenko managed to arrange an administrative ban of the conference a few days before its scheduled opening, simply by phoning Khrushchev.

When Khrushchev was deposed in 1964, Lysenko was deprived of his post as the director of the Institute of Genetics, and the institute was dissolved. This development would indicate that Lysenko had fallen from grace and yet when Zhores Medvedev published (via Columbia University) an English translation of his book on the history of the Lysenko affair,[168] he was dismissed from his job.

It was only after 1964 that Russian scientists were permitted to publish articles in which they could criticize the experiment with Jersey crosses. Voronov collated results from many farms that followed Lysenko's recommendations and bought pure bred bulls from Lysenko's farm. High butterfat was not retained in hybrids; it was proportional to the percentage of Jersey genes.[170] In another article Gorodinsky[171] wrote that Lysenko had exaggerated his butterfat figures by at least 0.29–0.49 per cent, and that the truth was that the yield of milk per cow dropped by 2660 litres per year! (Some large scale fudging!)

Throughout the following year the newspaper offensive against Lysenko continued, but neither he nor his supporters replied to the criticism. In 1965 Agranovsky visited Lysenko's experimental farm in Gorky Leninske, to study in detail the Jersey crosses. He came to the conclusion that the herd there was merely a showpiece and a result of concealed culling, the selected animals being put on a highly intensive feeding regime. At the end of 1965 the Academy of Science and the Soviet Ministry of Agriculture nominated a state review board. It uncovered a large number of data based on deceitful and fraudulent evaluation of experimental results. It also discovered that the experimental design of Lysenko's tests was not honest. The board concluded that Lysenko's methods were economically unsound, his recommendations erroneous and therefore recommended that all practices along these lines should be discontinued.

The fall of Khrushchev and of Lysenko led in 1965 to the reorganization of teaching in biology by culling Lysenkoist pseudoscience from it. It was another year before biology taught in Russian schools became similar to that prevalent in the West.

The story of Lysenko, and the victory of his pseudoscience over the well-established concepts of heredity in the twentieth century, is almost unbelievable. How was it possible for one man, and some of his opportunistic followers, to take hold of the reins of the entire agricultural establishment? By normal scientific standards it is indeed incomprehensible how Lysenko succeeded in rising to the top of Soviet agriculture, and how he managed to remove from his path all serious opposition from geneticists and other critics. If Lysenko's method had really succeeded, then it would perhaps be understandable. The Soviet attitude of intense commitment to practicality would accept practical results in agriculture, i.e. increased yield from crops, as sufficient argument to hail Lysenko as a hero. But were Lysenko's method successful?

I have already described several cases indicating that this was not so. Let us re-examine the story of potatoes. In 1936 instructions went to some 600 farms (tilling 18,000 hectares) to seed potatoes in summer as prescribed by Lysenko. When the harvest was in, a questionnaire was sent to all these farms. Only 420 replied and reported their results. Of these, Lysenko published as a proof of success the results of only the 50 best, covering altogether 407 hectares (about 2 per cent of the total area cultivated). These selected results were then used by the Council of People's Commissars as proof 'of the possibility of obtaining yields of non-degenerated potatoes twice as great as the usual spring planting in the southern part of Ukraine'. The experience of the unsuccessful farmers was simply not taken into account.

Another example of failure was the so-called cluster planting of trees, based on the assumption that tree seedlings cooperate rather than compete. In 1952 it became evident that of the millions of trees planted in Lysenko's clusters, more than half had died. By 1956 only 15 per cent were still alive.[172]

Lysenko and his supporters gained recognition for their ideas by 'distortion of facts, demagoguery, repression, obscurantism, slander, fabricated accusations, insulting name calling and physical elimination of opponents'. Thus, says Medvedev, 'for nearly 30 years the "progressive" nature of their scientific concepts was confirmed'.[168]

Lysenko's agricultural 'successes' were certified by political superiors and therefore served as conclusive evidence that Lysenko had discovered theoretical truth. This situation permitted Lysenko's collaborators and followers to publish fraudulent achievements, such as the transformation of rye into wheat, viruses into bacteria, plant into animal tissue etc. To substitute such claims the authors even used doctored photographs, showing, for instance, hazelnut limbs and pine spruce on hornbeam trees.[173]

Lysenko's biographer Joravsky uses a picturesque figure of speech: 'The Lysenkoites had forced political salts into the bowels of Soviet scientists, and some began to void themselves in public, the honorable one fouling themselves alone, the dishonorable trying to rub it off on others'.[159]

As long as Lysenko promulgated erroneous ideas and theories and supported them by experiments which, because of lack of better knowledge, were not well controlled, it was a legitimate activity. We have already shown (chapter 3) that even good scientists may at some time in their careers become involved in delusive research but because of this they do not necessarily become outcasts of science. In the case of Lysenko, however, he fanatically insisted on the correctness of his pseudoscientific ideas, completely ignoring the evidence to the contrary available in the Western literature. He often bolstered his experiments with manipulated data and finally brazenly used his influence in the political power structure of the Soviet Union to eliminate his opponents.

The desire and commitment of the establishment to obtain quick results, by simple, understandable methods, and to improve the agricultural crops (which were always lagging behind demand) resulted in impractical behaviour which not only set back Soviet agriculture by almost two decades, but did tremendous damage to the Soviet biological sciences.

6

Documented Cheating in
Clinical Research

Over the past 15 years or so several cases of scientific fraud have
come to light and received considerable publicity. This does not
mean that no misconduct in science had occurred before: I have
already described several incidences of alleged faking of experimental results at the beginning of this century (Kammerer, Carrel).
Because communications and ties between scientists and the public
were not as open and free as they are today, however, these 'frauds'
were discussed only in professional circles.

For the past few decades we have been living through a revolution
in science.[174] Newly created ideas are first circulated among specialists,
with varying success; they are then disseminated in professional
circles by correspondence, by circulation of preprints and by
publication. The 'revolution on paper' eventually becomes a revolution in science, when the new ideas are incorporated into the practice
of the discipline.[175] In biological sciences alone we have witnessed a
number of epoch-making discoveries, such as the double helix of DNA
by Watson and Crick. The unravelling of the cellular molecular process led to an explosion of genetic engineering. The discoveries of a
multitude of antibiotics, the deepening understanding on neuronal
and hormonal communication in animals are also a part of this
revolution.

Until the second third of this century medicine was mainly a
practical art of physicians. Their success depended on practical
experience passed on by their predecessors and accumulated during
their lifetime. Medicine has now entered a new phase, becoming a
true scientific discipline in all aspects of physiology, endocrinology,
pharmacology and surgery.

I believe that the sudden accumulation of reports of cheating in science may be attributed to the increased impact of science on practically every aspect of our lives and the public awareness of the interaction between science, governments, industry and academic institutions. This century has witnessed an enormous increase in the number of practising scientists. It is not an exaggeration to say that more scientists live now than have lived before throughout human history. Cheating among such numbers, even if marginal, becomes more visible.

THE SUMMERLIN AFFAIR

The Summerlin affair involved the falsification of results of skin transplantation in mice by a doctor/ scientist, William T. Summerlin.

It is generally known that transplantation of organs from one individual (donor) to another (recipient) has generally been unsuccessful unless the donor and the recipient are genetically identical (the same strain of inbred animals) or are identical twins (humans). During evolution, the immune system has developed so that it has became finely tuned to minor differences in the antigenic composition of the 'self'. The surface of each cell in the organism is equipped with histocompatibility antigens (HLA), i.e. complicated glycoproteins, the structure of which is genetically determined. Even small differences in this structure are recognized by the immune system (by the lymphocytes bearing the name 'killer cells') and these foreign, intruder cells are eliminated.

The present success of organ transplantation (heart, kidneys etc.) is based on steps taken to suppress the immune response of the recipient against the grafted organ. This success is also ensured by careful typing of the HLA and by using organs and tissues that show the least differences from those of the recipient. Since the need for tissue and organ transplantation in modern society is great, enormous efforts have been made by the scientific community to find ways of successfully grafting organs, including skin.

Summerlin's approach was based on the assumption that pieces of skin of mice, removed from a donor and held for appropriate periods of time in nutrient solutions outside the donor's body, would lose some of their HLA; this would make the cells in the tissue less vulnerable to immune surveillance mechanisms of the recipient and thus make the graft acceptable. Summerlin had claimed success in this type of experiment before joining the prestigious Sloan Kettering Institute in New York, but it was there that his fraud was detected.

Summerlin had been experimenting on transplantation of skin from black mice to white ones. On the morning of 26 March 1974, while an elevator was carrying him and a cage containing animals with transplanted skin patches from his laboratory to the office of the Director of the Sloan Kettering Institute, Dr Robert Good, Summerlin used his black felt-tip pen to darken an area of transplanted graft on two white mice.

The falsification was exposed the same day when, after the demonstration in the director's office, the mice were brought back to the animal room. Senior laboratory assistant James Martin noticed that the appearance of the black grafts on the white mice was unusual; he found that the black colour could be washed away with alcohol, and reported his findings to a senior technician, who in turn brought the matter to the attention of a research fellow, and through him to Dr Lloyd Old, vice-president of the Institute, and to Dr Good. By noon Summerlin was summoned to the director's office where he admitted his wrong doing, upon which Dr Good suspended him temporarily from all administrative and scientific activities. A few days later Good appointed a peer review committee, composed of five senior scientists from the Institute, to investigate the matter.

The committee studied the case in depth and produced a report on 20 May 1974. The committee confirmed that Summerlin had misrepresented his findings and observations, and also came to the conclusion that there were other irregularities in his scientific career and achievements: 'The committee members believe that some actions of Dr Summerlin over a considerable period of time were not those of a responsible scientist'. His 'irresponsible conduct is incompatible with discharge of his responsibilities in the scientific community'. The committee therefore recommended that Summerlin be offered a medical leave of absence to alleviate the situation. How had this extraordinary situation come about?

William T. Summerlin was raised in South California, went to Emory University and graduated from the Medical School there in 1964. After a period of surgical–medical internship at the University of Texas Medical Branch in Galveston, he served for two years at Brooke Army Center at Fort Sam Houston. In 1967 Summerlin became a resident in dermatology at Stanford, and by 1970 he was chief of dermatology at Palo Alto Veteran Hospital. In August 1971 he joined Dr Robert Good in his immunology laboratory at the University of Minnesota, and when Good accepted the directorship of the Sloan Kettering Institute in the spring of 1973, Summerlin moved there too, first as a visiting investigator. Later he was appointed as a member of the Institute, a rank equivalent to a professorship.

During the period 1967–73 Summerlin engaged in well-publicized research on prevention of graft rejection. According to his hypothesis, placing the donor organ in tissue culture for about a week before transplantation would prevent the subsequent rejection of the graft. The theoretical explanation of why this should be so was based on the assumption that the *in vitro* cultivation of the tissue to be grafted would deplete the tissue of indigenous white blood cells (lymphocytes), and would cause the loss of some of the molecular structures from the surface of the cells which identify them as foreign to the recipient. According to Summerlin, when such tissue was transplanted there was no infiltration of the grafts by the host lymphocytes. Summerlin published accounts of his successful results with grafts of cultured skin from unrelated and HLA-incompatible human donors; also of skin from white mice to C3H dark grey ones, of human corneas into rabbit eyes, and of adrenal glands from one strain of mice to another.

The Institute investigating committee found, however, that the past successes of Summerlin were doubtful. The only surviving mouse with transplanted skin which Summerlin brought to New York from Minnesota ('Old Man') turned out to be a descendant of a cross between C3H (recipient of the graft) and the white A strain (donor) (F_1 hybrid). Since such a hybrid animal would accept grafts from both strains of mice involved in the genetic crossing, it was not surprising that the white A graft survived well on the C3H mouse. The claim that there was a successful transplantation of skin between two different pure-bred strains of mice was, of course, unfounded. The committee found it

> difficult to credit that Dr Summerlin could have been continuously ignorant of the elementary possibility which became even more critical as this mouse assumed the role of a single exceptional success, or that he remained unaware that simple tests of any of several colleagues would have immediately settled the matter.

This evaluation of the committee threw doubts on the other 'successful' transplantations Summerlin had done in Minnesota. All mice involved in the Minnesota experiments were dead, and therefore Summerlin's claims could not be confirmed. When asked by the committee why the successful grafted recipients were not retained for observation, Dr Summerlin gave 'the astonishing reply that they were sacrificed at intervals to provide serum to be stored for H-2 antibody tests at a later date. It is scarcely conceivable that Dr Summerlin, in such an immunologically sophisticated environment as Dr Good's group in Minnesota, would have believed that it was necessary to kill a mouse to obtain a serum.'

The committee also found that the data on grafts of tissue transplanted from one species to another (as for instance from guinea-pigs, rabbits and pigs to each other) were at the best vague. While examining a paper submitted by Summerlin to the *Journal of Experimental Medicine*, the committee noticed that the percentage figure for acceptance of mouse skin allografts (grafts within the same species) did not correspond to the number of animals cited in a table elsewhere in the paper. The paper was withheld from publication upon the request of Dr Good, whose name also appeared among the authors.

The committee also stated that investigators had been justified in

assuming that pronouncements of new observations by other scientists are made on the basis of adequate data, systematically obtained, recorded and interpreted according to accepted standards of investigation; since this is evidently not the case with Dr Summerlin's experiments with mice, it is the opinion of the committee that he has propagated misconceptions concerning the nature and scope of his work in this area.

In the late 1960s Summerlin had been involved in research into corneal grafting. Although quite commonplace today, the mechanism of acceptance and rejection of the grafts had not been completely elucidated at that time. While in Minnesota, Summerlin worked on corneal grafting in rabbits in a team composed of Drs George E. Miller, D. Doughman and John E. Harris from the department of ophthalmology. In one experiment, pieces of cultured and non-cultured (fresh) cornea were put in a pocket behind the eyeball to test for inflammatory reaction. In another experiment, central corneal transplantations were made, namely, a piece of guinea-pig, human, chicken or rabbit cornea was transplanted to replace an excised cornea in the recipient rabbit. Reporting upon these experiments, Summerlin stated:

allotransplants and xenotransplants of cornea near limbus (the edge of the conjunctiva that overlays the cornea) are accepted and have exhibited good function for 6 months. Control allogeneic and xenogeneic grafts placed near limbus are regularly rejected within 2 weeks.[176]

The investigating committee, however, came to the conclusion that corneal xenografts were, in fact, unsuccessful by both techniques, and that no xenografts survived for more than 40 days, though Summerlin claimed that they had survived for six months.

Summerlin continued his work on corneal transplantation in New York with the cooperation of Drs Peter Laino and Bartley Mondino, both ophthalmologists at the New York Hospital. They used corneas from fresh human cadavers and transplanted them to rabbits. Paired – one fresh and one cultured – human corneas from the same donor were to be transplanted into both eyes of a recipient rabbit, one eye receiving the untreated cornea and the other the cultured human cornea. As the investigation showed, however, the transplantations were made so that the two corneas were grafted each into one eye of a different rabbit, the other eye remaining as an untreated control. Drs Laino and Mondino transplanted some 20 rabbits with human donor corneas which had first passed through tissue culture. 'It appears that no appreciable difference in acceptance or rejection time occurs when these eyes are compared to others grafted directly into recipients without having first been passed through tissue culture. They both fail about the same time' (letter of Dr Laino to Summerlin 25 February 1974). As a result of these discouraging results, the ophthalmologists abandoned the work in December 1973.

Summerlin, however, claimed otherwise. He 'assumed' that the ophthalmologists followed the planned protocol of preparing a double corneal graft and in October 1973 he displayed the operated rabbits at a meeting of the scientific consultants of the Sloan Kettering Institute (Sir Peter Medawar and Dr Stock) in the presence of Dr Good. Summerlin showed the consultants a rabbit with one operated eye where the cornea had become cloudy, and stated that this was the eye into which a fresh cornea had been transplanted; the other, perfectly clear eye was exhibited as the one receiving the cultured cornea.

Medawar remarked about this incident:

> Through a perfectly transparent eye this rabbit looked at the board with the candid and unwavering gaze of which a rabbit with an absolutely clean conscience is capable. I could not believe that this rabbit had received a graft of any kind, not so much because of the

perfect transparency of the cornea as because the pattern of blood vessels in the ring around the cornea was in no way disturbed. Nevertheless I simply lacked the moral courage to say at the time that I thought we were victims of a hoax or confidence trick.[177]

Dr Nineman, a research fellow in Summerlin's laboratory, had some doubts about the story that Summerlin had told the board while showing them the operated rabbit. He therefore contacted Dr Laino and found out from him that indeed the grafts were done into one eye only. He told Summerlin that his description of the rabbit had been in error and Summerlin accepted this criticism as justified. Nevertheless, he kept on, not only presenting the rabbits as having double corneal grafts on several occasions in November and January, but also supplied for publication a photograph of a rabbit said to have a double eye graft (*Medical World News*, 15 March 1974) when in fact one eye was not operated at all.

Taking all these findings into account, the Sloan Kettering Institute committee concluded that 'Dr Summerlin was responsible for initiating and perpetuating a profound and serious misrepresentation about the results of transplanting human corneas in rabbits.'

The committee's report did not censure Summerlin alone. It also referred to the role played by Dr Good, the director of the Institute:

> . . . it would appear that the great demands made upon Dr Good, especially after he took over the directorship, compromised his ability to personally supervise projects of Dr Summerlin that conspicuously lacked the sound experimental planning and guidance that Dr Good could have provided in another circumstance . . . The committee feels that Dr Good shares some of the responsibility for what many see as undue publicity surrounding Dr Summerlin's claims, unsupported as they were by adequate authenticated data.

I concur with the Committee's conclusion that a director of a laboratory who also is a collaborator in research should have had his junior colleague more closely supervised. Soon after the Summerlin affair blew over, Dr Good relinquished his post at the Sloan Kettering Institute, and is now at the School of Medicine of the University of Florida in St Petersburg.

Dr Lewis Thomas, president of the Memorial Sloan Kettering Cancer Center thought that the most rational explanation for Summerlin's performance was that he had been suffering from emotional disturbance such that he had not been fully responsible for his actions. The committee supported this view and explained Summerlin's behaviour as due to self-deception or some other aberration, which hindered him from adequately assessing the consequences of his conduct. Nevertheless, the committee also appreciated Summerlin's personal qualities of warmth and enthusiasm and his confidence in himself, which was persuasive enough to convince others about the correctness and importance of his findings. These were the reasons that had led Dr Good to appoint Summerlin as a Member of the Institute and Laboratory Head.

When the affair of the painted mouse exploded Summerlin gave a statement to the press (28 May 1974) in which he said (as reported by McBride[178]):

> My error was not in knowingly promulgating false data, but rather in succumbing to extreme pressure placed on me by the Institute's director to publicize information regarding the rabbits, information which I informed him was best known to ophthalmologists, and to an unbearable clinical and experimental load which numbed my better judgement to consult with ophthalmologists, rather than to rely on my assumption prior to making my statement.

In addition to explaining the situation by personal and professional stress, mental and physical exhaustion, Summerlin tried to move part of the blame onto his superior (Dr Good), by stating:

> . . . As the youngest member of the Institute, I was charged with the responsibility of heading a laboratory in the Institute, while serving as head of clinical service at Memorial Hospital. Within my lab there were 25 separate research projects being conducted, of which the corneal study represented only an ancillary undertaking. Further, I was personally engaged in 26 collaborative efforts with scientists in ten countries. My clinical load averaged 6 hours out of a day that usually began at 5 a.m., and on days when I did not sleep in the lab, ended at 6 p.m. . . . Secondly this personal pressure generated by my schedule was aggravated by the professional pressure which is regrettably so much part of medical research. Time after time, I was called upon to publicize experimental data and to prepare applications for grants from public and private sources. There came a time in the fall of 1973 when I had no new startling discovery, and was brutally told by Dr Good that I was a failure in producing significant work. I was placed under extreme pressure to produce.

These pressures, said Summerlin, had led to complete mental exhaustion. He ended his statement to the press by expressing the hope that other dedicated young researchers would not fall into the same pattern which brought about his downfall. The moral Gail

McBride draws from Summerlin's story is that gifted men should not grow so busy that they cannot find enough time for truly important matters.[178]

Summerlin's story has been told in great detail in Hixson's book *The Patchwork Mouse*.[179]

We now have to ask the question, Who should be blamed for this fraudulent behaviour? Should it be only the deviant individual, or should the whole system be accused?

When deviant scientists, like Summerlin, published their findings, these were accepted as genuine because they provided the 'right' and expected answers, and because their colleagues, supervisors and reviewers wished to believe their findings. We have seen this happen in cases discussed in earlier chapters. More than this, however, a head of a laboratory will obviously prefer to have a scientist who steadily provides good, clean, publishable data, rather than a cautious one who is not easily satisfied with the results, repeats his experiments again and again and is worried by an occasional lack of success.

So the pressure is on, and many young and ambitious scientists working in large teams may directly feel it, even in as casual a greeting from the head of his laboratory as 'What's new' today?'. In the case of Summerlin, even without any pressure from above, the 25 research projects and the 26 collaborative studies that were under his supervision might have been sufficient to precipitate his breakdown. In addition, the pressure Summerlin was speaking about was the perception of what had been expected of him by the director whose financial and moral support was important for his success.

The treatment of the Summerlin affair by the lay press and even by *Science* has been castigated by June Goodfield:[180] '. . . the press forgot the basic ethics of reporting and the professional standards of their jobs.' The only carefully investigated and researched report concerning the Summerlin affair was an article by Gail McBride, associate editor of the *Journal of the American Medical Association*,[178] in which 'blame and understanding, sympathy and fair judgement were most fairly handed out to all parties . . . and something approaching truth was out at last'[180]. Other reports in the press abandoned one ethical principle – they did not dig hard enough to uncover the truth and do justice with professional competence. They decided a priori that the case was one of cover-up, based on presumption of guilt.

THE BRILLIANT JOHN DARSEE

The Summerlin affair at Sloan Kettering in 1974 virtually repeated itself in 1981 at Harvard. Here a brilliant and promising researcher and physician, John Darsee, was caught faking results in the laboratory of a prominent research leader and clinician.

Dr Eugene Braunwald, an eminent cardiologist, trusted his young collaborator Darsee, and lent his name to at least five papers produced during Darsee's 15 months of work in Braunwald's laboratory. As in the case of Summerlin, when a faking incident came to light a thorough investigation, at Harvard, the National Institutes of Health and Emory University where Darsee had worked, revealed that the incident was not an isolated one. Eight of ten papers that Darsee had published before joining Harvard had to be withdrawn or corrected. One paper was found to be completely fictitious, and out of 45 abstracts of papers presented at various conferences only two stood up to scrutiny.[181]

John R. Darsee studied at the Notre Dame University in Indiana (USA) and received his medical degree in 1974 from Emory University in Atlanta (as, coincidentally, had Summerlin.). Having excelled in clinical work and research at Emory during the following five years, Darsee moved to Boston in July 1979 to join Eugene Braunwald's laboratory studying cardiac output. Braunwald had come to Harvard in 1972 as Professor of Theory and Practice of Physic, and later became physician in chief of Peter Bent Brigham and Women's Hospital. In 1980 the Brigham department was joined with that of Beth Israel Hospital.

Shortly after arriving in Boston, Darsee, working on an NIH fellowship, was made an instructor at Harvard Medical School at the Department of Physiology. He was scheduled to obtain a faculty appointment on 2 July 1981. The laboratory in which Darsee conducted his research was headed by Dr Robert Kloner. The team comprised one other scientist and two technicians. Darsee tested various drugs on dogs with an artificial myocardial infarct: the drugs were designed to prevent or diminish the damage to the heart muscle following the blockage of the blood supply to it.

In May 1981 Darsee's co-workers approached Dr Kloner with a problem: they suspected that an abstract Darsee was preparing to be sent off for publication had no factual basis, in fact, it seemed to be based on experiments that had not been done. Kloner asked Darsee to provide him with the raw data connected with this research, such as electronic recordings of heart activity, blood flow measurements, tissue sections of hearts of dogs etc. This experimental series was supposed to have lasted for several weeks. Darsee, however, did not have these data. In order to comply with Kloner's request, Darsee set up an experiment on 21 May. He hooked up one dog to the apparatus that measured the blood flow and electrical activity, treated the animal with the drugs under study and started taking readings. To the utter amazement of the co-workers who were present, Darsee marked the chart coming out of the machine as day one, day two and so on, as if the data were coming out of an experiment lasting for two weeks. This manipulation was immediately reported to Kloner and when confronted with the eyewitnesses, Darsee admitted his wrongdoing, but explained that he was only reproducing tracings from previous experiments which he had unfortunately lost. He also said this incident had been a single foolish act of misconduct.

Kloner reported Darsee's misdeed to Braunwald, who immediately started an investigation into Darsee's research. When he discovered further irregularities, Braunwald terminated Darsee's clinical and academic appointments at Brigham and Women's Hospital, as well as his NIH fellowship, but he did not inform the NIH authorities of the reasons of this termination. Darsee was permitted to continue his work in the laboratory, only now under closer supervision.

In spite of these events, Darsee was permitted to submit ten abstracts to the journal *Circulation* (six of them were from Harvard, and five were co-authored with Kloner and Braunwald). On 31 August Braunwald presented to the National Academy of Science a paper authored by several workers from his laboratory, including Darsee.[182] In October 1981 the *American Journal of Physiology* printed a further paper in this series,[183] and still another article, authored by Darsee and Kloner,[184] appeared in the *American Journal of Cardiology*. The publication of these papers attests to Braunwald's belief that Darsee's misconduct on 21 May was an isolated and insignificant incident, not connected with any other research, either past or ongoing.

In the meantime, Darsee was working under close supervision on a multi-institutional study sponsored by NIH. This study, an 'Animal model of protecting ischaemic myocardium', was designed to find out whether certain drugs could limit the damage occurring in the heart during a heart attack. The other laboratories participating in these studies were from Duke University, VA Medical Center and Johns Hopkins University. In September 1981 Braunwald's laboratory submitted its coded data of this study to the NIH. The NIH panel that collated the results of participating laboratories discovered serious discrepancies between the data submitted by Harvard and those of other laboratories. The Harvard data, collected by Darsee, indicated that the drugs ibuprofen and verapamil had limited the damage to the heart muscle when the coronary artery was blocked. These findings contradicted the results of collaborating laboratories (Figure 6).

When challenged, Darsee could not substantiate the measurements of blood flow which he had performed prior to 21 May. When this fact became public knowledge, the Dean of the Harvard Medical School, Daniel C. Tosteson, appointed in November 1981 an ad hoc committee, composed of eight professors from Johns Hopkins, New York and Harvard under the chairmanship of Dr Richard S. Ross, Dean of the Faculty of Medicine of the Johns Hopkins School of Medicine, to investigate the affair. Independently, NIH appointed a panel of four members, chaired by Howard E. Morgan of the Penn State University College of Medicine, to examine the implication of Darsee's data. The Ross committee issued a report in January 1982 (*Harvard Gazette*, 29 January 1982), and the Morgan panel at the end of March 1982.[185]

The Harvard committee found that Darsee's study of the size of myocardial infarction as measured by the disappearance of a

| | Collaborating laboratories | | | Darsee's data | |
	1	2	3	before 21 May	after 21 May
Myocardial blood flow Standard deviation	0.28	0.38	0.25	0.04–0.05	0.31
Mean heart weight (gms)	84.8	87.8	71.2	141.9	83.1
Ratio heart weight (gms) to dog weight (kg)	4.42	4.36	5.07	7.44	5.01

Figure 6. Darsee's data from the collaborative NIH study, before and after his fraud in another experiment was exposed.

phosphorylating enzyme from the blood 'appears to have been manipulated'. This came to light when Darsee's data were compared to those collected by Dr Leonard Holman of the Nuclear Medicine laboratory. Holman's data were unbelievably close to those of Darsee.[186] The investigation also revealed that analysis of heart tissue of dogs that should have been injected with radioactive material showed no radioactivity prior to 21 May, but that the tissue of dogs operated on after this date was radioactive. Since in order to measure radioactivity in the heart's muscle the hearts had to be removed from the dogs' cadavers, the experimental dogs should have been buried with their hearts missing. In one case the exhumed dog was found to have been buried with its heart intact.

The Morgan panel uncovered 'extensive irregularities' in five papers authored by Darsee, Kloner and Braunwald. Analysing the circumstances leading to the fraud, the panel stated:

> The panel is of the opinion that the circumstances prevailing throughout Dr Darsee's period in the laboratory, while not responsible for or in any way condoning his misdeeds, helped to create an environment that may have inhibited their being uncovered . . . there just wasn't enough direct contact with Braunwald.[181]

Since this statement could have been understood as a reprimand to the director, Braunwald sent a memorandum to the NIH stating:

> Since 1960 I have maintained my direct participation in the laboratory by trying to offer these [leadership, ideas, inspiration] while at the same time providing an extremely capable colleague, such as Dr Kloner, for day-to-day, on site fulltime participation in actual experiments.

The Darsee matter did not stop there. In April 1982 Braunwald reviewed some of Darsee's papers from Emory and was concerned about their integrity. He reported his findings to the NIH and to the chairman of the Internal Review Panel at Emory University, Neil Moran. The Emory investigation of Darsee's activities ended there with Moran's statement on 5 May 1983 that only two of the ten papers Darsee had published from Emory were considered valid, and out of the 45 abstracts, only two stood up to scrutiny, and many of them appeared to be fictitious.[181] Moran's investigation also detected that a paper by Darsee and Heymsfield[187] was probably fictitious since it did not identify the hospital where the study had been done, or the patients involved. Moreover, the paper ended with the following acknowledgement:

> We are indebted to Johnson S. Caulder, PhD for performing the amino acid analyses, to Lawrence D. Bergmann, MD for obtaining the heart

tissue during pacemaker placement, to Myron C. Filstein, MD for referring several family members.

The persons whose work was acknowledged in this statement were found not to exist.

Possible repercussions of the Darsee affair are discussed by Culliton.[181] Darsee left Harvard and obtained a clinical position at Ellis Hospital in Schenectady, NY. In spite of all the adverse publicity he had received, the hospital staff thought him a brilliant and charismatic doctor. From the Ellis Hospital Darsee issued this statement: 'I am asking for forgiveness for whatever I have done wrong and want to contribute to the medical system'[181] (quoted by Broad in *New York Times*, February 1983, A1, A23).

In a letter to Braunwald written in December 1981 (quoted in the *Journal of the American Medical Association* of 8 April 1983 p. 1806), Darsee excuses his behaviour:

> This was an extremely difficult period for me. I had too much to do, too little time to do it in, and was greatly fatigued mentally and almost childlike emotionally. I had not taken a vacation, sick day, or even a day off from work for six years. I had put myself on a track that I hoped would allow me to have a wonderful academic job and I knew I had to work very hard for it.

Why was Darsee's misconduct, stretching over a period of almost seven years, not detected earlier? 'It takes a different mind set to look at a paper and think it total fraud', remarked Morgan of NIH. Collaboration in scientific research is based on mutual trust and confidence in the integrity of all participants of a research group; a well-designed fraud will always take everybody by surprise.[188-190]

Another question for which I have no ready answer is should any blame have been attached to Braunwald, who after all had his name included in the fraudulent papers? Is it satisfactory merely to state that there should be no penalty for absolute trust in scientific research done by a colleague or student thought reliable, even when it is revealed that that trust has been abused?

PUBLISH AND BE DAMNED

In 1975, Indian-born Dr Vijay Soman obtained a fellowship at the Yale School of Medicine in the laboratory of Dr Philip Felig, Professor of Endocrinology. There Soman engaged in the study of insulin metabolism. In 1977 he became an associate professor at Yale.

In February 1980 Soman found himself facing an auditor in connection with an accusation that he had used false data, and that a research paper submitted by him for publication, with Felig as co-author, contained plagiarism. The auditor in this case was the chief of the diabetes metabolism unit at Beth Israel Hospital in Boston, Dr Jeffrey Flier.

What was the basis of this accusation? The ball had started rolling a year earlier, when Dr Helena Wachslicht-Rodbard, an endocrinologist at the National Institutes of Health, complained to her supervisor, Dr Jesse Roth, that she had been plagiarized by Soman and Felig[191] in their paper on insulin binding in anorexia nervosa (a nervous condition in young adults characterized by complete loss of appetite). This paper, submitted by Soman and Felig to the *American Journal of Medicine*, had been sent to Roth for review. He passed it on to Dr Wachslicht-Rodbard, who in reading the manuscript discovered that at least 60 words in it were an exact copy of a similar passage in her own manuscript, co-authored with Roth, and submitted for publication in November 1978 to the *New England Journal of Medicine*.[192] That manuscript had been returned by the editor, Arnold Relman 'because of differences of opinions among our referees'. Dr Wachslicht-Rodbard became very upset, because she correctly assumed that not only had Soman and Felig seen her manuscript through the editorial office of *New England Journal of Medicine before* they had sent their own manuscript to the *American Journal of Medicine*, but that they also were the referees who had advised Relman not to publish her paper.

The papers in question dealt with the problem of binding of insulin to monocytes in the blood of patients with anorexia nervosa and how this binding changed with progressive cure. As a result of Roth's inquiries about the rejection of the paper, Relman asked for clarification from Felig, to whom he had sent the manuscript for review. Felig had not thought it necessary for him to disqualify himself as a referee, in spite of the fact that he had been working with Soman on a very similar problem. It can be said, after all, that one wants referees to be experts on the subject of research they are requested to judge.[193, 194]

When Relman found out from Wachslicht-Rodbard that Soman and Felig had submitted a manuscript on the same topic, he called Felig to express his concern. So also did Roth, who had been a boyhood friend of Felig. Felig assured both inquirers that the work had been done independently, but when shown by Roth the identical passages in the two manuscripts, agreed to make certain changes. He also agreed not to publish his own paper, co-authored with Soman, before that of Wachslicht-Rodbard appeared in print. Roth accepted this arrangement, but Wachslicht-Rodbard, thinking this was a cover-up, wrote to Dr Robert W. Berliner, the Dean of the Yale Medical School, in March 1979 accusing Soman and Felig of plagiarism. She also stated that the data in Soman and Felig's paper were improbable, first, because of the claim made there that anorexic females who did not menstruate while anorexic resumed their periods upon gaining weight (which she knew was untrue), and secondly that the insulin binding data were too close to an ideal curve (a statistically improbable fact). In short, she claimed that the data were fudged.

In the meantime Felig confronted Soman, who admitted making a copy of Wachslicht-Rodbard's manuscript and to copying certain phrases from there. He also confessed to having used her equations to calculate his own data. Felig reprimanded Soman for these infringements of scientific ethics, but still believed Soman to be trustworthy and honest.

Upon receipt of Wachslicht-Rodbard's complaint, Dr Berliner asked Felig to verify the data used in his paper. Soman showed Felig a list of six patients he had worked with (but not hospital charts) as well as average data he had compiled during the study (but not hospital charts) as well as average data he had compiled during the study (but not the raw data). This seems to have satisfied Felig, and he informed Berliner that the work of Soman had been done as reported in the manuscript. Berliner passed on this information to Roth and Wachslicht-Rodbard, but she would not accept this explanation and hinted to Roth that at the next meeting of the American Federation of Clinical Science she would publicly denounce Soman and Felig, unless the matter was thoroughly investigated. Facing this situation Felig and Roth agreed to call in an auditor. Regrettably the person chosen allowed eight months to pass without starting the investigation.

In January 1980 the challenged paper by Soman and Felig appeared in print,[191] in spite of Felig's promise that it would not be published before the matter had been settled.

Wachslicht-Rodbard was furious and demanded immediate action from Roth. Thus, another auditor, Dr Jeffrey Flier, was selected. He went to Yale, met Soman and inspected his data. He was shown the records of only five patients (out of the six reported in the paper), all with anorexia nervosa. When asked about the insulin binding data, Soman produced for each patient only sheets of raw numbers; when asked about the graphs he should have made for each patient he is reported to have said: 'Well, we threw away the individual graphs after a year because we had no storage space.' Flier then inspected the raw numbers and found that they did not correlate with the data presented in the published paper. 'There was no way that the beautiful composite curves they had in the paper could have been derived from the data I had been looking at' (quoted by Hunt in *New York Times* Magazine, 1 November 1981, p. 42). When confronted by Flier with the evidence for mismatch between his data and those in the paper, Soman admitted fudging and explained he had been under pressure to publish as fast as possible so as to obtain priority for the funding.

Dr Flier, speaking about the confrontation said 'Although some doctoring of data had gone on in this laboratory in the past he [Soman] felt that this was not significantly different from what he believed to go on routinely elsewhere.'

On 12 February 1980 Flier informed Felig that Soman had falsified data. Felig conferred with Berliner and they informed Dr Samuel Thier, chairman of the Department of Internal Medicine. Thier confronted Soman; who admitted having forged his data. He was asked to leave Yale and in April he left the USA and returned to Poona, his home town in India.

Felig, Berliner and Thier impounded all the records made by Soman to find out whether any of the other 14 papers previously published from the laboratory for which Soman had supplied data were fraudulent. Dr Jerald Olefsky, an endocrinologist of the University of Colorado, was asked to investigate the raw data and the papers.

On studying the records, Olefsky found that at least one-quarter to one-half of all the data relating to published material had been discarded and were missing. Most of the published conclusions were qualitatively correct, but in the end only two out of the 14 papers inspected were above suspicion. The rest either contained falsified data or were of questionable significance because the supporting data were missing. In May and June Felig withdrew these 12 questionable papers; in eight of them he was a co-author.

The discovery of Soman's fraud had come at a very inopportune time for Felig. He was being favourably considered as a candidate for the position of Samuel Bard Professor and chairman of the

Department of Medicine at the College of Physicians and Surgeons (P&S) of Columbia University. In May 1980 Felig was invited to meet the senior advisory faculty at the P&S in connection with the forthcoming appointment. Speaking about this meeting, Morton Hunt of the *New York Times* wrote:

> The faculty members present later attested during an inquiry that they heard nothing from Felig about the plagiarism issue, or about the failure to discover, in the course of a year, what Flier had been able to find in three hours. (*New York Times* Magazine, 1 November 1981)

Felig began his new duties at Columbia in June. At the end of July 1980 Felig presented the Dean of Columbia Medical School Faculty with all the documents pertaining to the Soman affair. A committee of six senior faculty members discussed the matter for four days and in the end advised Felig to resign because of 'his poor judgement if not negligence' and came to the following conclusion.

> The committee must conclude that the events disclosed by the correspondence, and Felig's attitude when asked about these events, reflect ethical insensitivity and the application of unacceptable standards to scientific research. (Altman, *New York Times*, 9 August 1980, p. B660)

With deep regret the committee decided that Felig should not retain his professorship and posts at Columbia. On 5 August, Felig resigned.

How did the story end? Dr Helena Wachslicht-Rodbard resigned from the NIH and from research and entered private medical practice. Dr Felig was reappointed by Yale as tenured professor in November 1980. Six months later, in March 1981, Felig appeared before the Subcommittee on Investigations and Oversight of the Committee on Science and Technology of the US House of Representatives.[195] The session, chaired by Albert Gore Jnr., was devoted to fraud in biomedical research.

Felig testified at this session that he had made several misjudgements in the case of Soman. First, that he accepted Soman's lists of patients, instead of examining the original patients' charts. Secondly, that he waited for eight months for an outside auditor instead of instituting an internal audit. Thirdly, that he proceeded with the publication of the paper without waiting for the resolution of the alleged fraud.

Analysing the reasons for these misjudgements, Felig described the situation in the laboratories. In some laboratories senior scientists review all original data collected by their junior colleagues; in other laboratories the senior scientists limit their interaction to general discussion of the outline of the problem to be investigated. A delicate situation develops when the junior scientist comes up with a new technique with which the senior scientist is not directly familiar. In such a case it is necessary for the senior scientists either to learn and become accustomed to the new technique so as to be able properly to judge its potential and limitations, or, alternatively, he has to trust his junior when putting his name on a joint publication. 'When a senior scientist is not too familiar, he or she should exercise even greater care in reviewing original data of the junior scientist, or his or her name should not be included on the paper'.[195] Felig further emphasized that 'collaborative research involves a relationship between investigators based on mutual trust' and on the assumption that the partners are honest.

The ultimate responsibility for preventing and uncovering data falsification *must lie* with the senior scientist and senior administrative officials in the research institution. This responsibility is shared by the scientific community as a whole by its commitment to replicate and await confirmation of new data.

Concluding his testimony before the subcommittee, Felig felt that although most scientists work long hours, give up leisure time and some financial advantage of non-academic career

> the desire for success may in some individuals override the principles of professional ethics . . . it is my belief that it is not the system of academic advancement (pressure to publish or perish) which inherently influences the use or interpretation of research data. Where such misuse or misrepresentation does occur, it is the unfortunate reaction on the part of some individuals to that system . . . it is clear to me that the senior scientist should constantly be examining his or her behavior vis a vis the junior colleague to avoid actual or perceived pressures.

Should the last word go perhaps to Dr Rehnie of the *New England Journal of Medicine*? He wrote that, in spite of all, nothing should justify plagiarism: 'All this garbage about the pressure of science is ridiculous. Scientists have no more or no less pressure than coal miners or sailors or the rest of the World' (quoted by Roark[196]).

A MISDEED REVEALED BY A LOGBOOK

Between the years 1973 and 1977, Dr John Long, in collaboration with others, published a series of articles showing how malignant cells taken from patients with Hodgkin's disease could be grown in laboratory glassware and repeatedly 'subcultured'.[197-200] The availability of such permanent lines of tumour cells was important in the study of this malignancy. (The term 'permanent lines' relates to the ability of cells in culture to multiply continuously, an ability, as already discussed in chapter 4, which normal, non-cancerous cells do not possess.)

Long came to the Massachusetts General Hospital in Boston in 1970. In 1972 he joined the laboratory of Dr Paul Zamecnik, and in 1979 he was promoted to associate professor.

In 1979 there arose some doubts about the identity of the Hodgkin's cell lines established by Long and his associates, as a result of an investigation related to a paper published by Long in 1977.[201] About a year before this paper was published, two collaborators from Zamecnik's laboratory, and two collaborators from the Electron Microscopy Unit, prepared a paper for submission to the *Journal of the National Cancer Institute*. This paper dealt with the presence in sera of patients with Hodgkin's disease of antibodies directed against cells in tissue cultures derived from a Hodgkin's tumour.[199] The properties such as the size of the serum components reacting with these cells was estimated on the basis of data from sucrose gradient centrifugation. In this technique, sucrose solutions of decreasing concentrations are layered in centrifuge tubes, from bottom to top, and the test substance (the patients' sera) is deposited at the top of the gradient. The tubes are then centrifuged for a suitable time period, at the end of which components of serum of varying densities 'settle' in regions of sucrose of equivalent density. The paper was returned by the editor with some remarks concerning the size of the complexes, since they had seemed to the referee to be much smaller than expected. The referees' comments were discussed by Long and S. C. Quay, who had collaborated on the project. The next day Quay left Boston for a two-weeks vacation; when he came back, Long told him that in his absence he had repeated the centrifugation experiment and obtained a result fitting well with the expected values mentioned by the referee. The manuscript was corrected accordingly, dispatched to the editorial office, accepted for publication in October 1978, and published in April 1979.[199]

In the autumn of 1979 Quay reinvestigated the problem of the size of the complexes and asked Long for his data from the experiment he had carried out the previous year, when Quay had been on vacation. Quay inspected the notebook and had the impression that the data were not based on a real experiment. He expressed his doubts to the chairman of the Department of Pathology, Dr Robert McCluskey. Confronted with this suspicion, Long produced the ultracentrifuge logbook to prove that he had actually performed the experiment. (According to a common procedure in laboratories each run of an ultracentrifuge is recorded in a logbook with details of the operator, the date, the type of rotor used, the speed of revolutions as well as the initial and the final numbers of revolutions made by the rotor during the run.) Quay computed from the logbook that in Long's disputed experiment the run was much shorter than that required for a proper separation of the immune complexes in a sucrose density gradient. When shown this calculation, Long admitted that he had contrived the results.

On 31 January 1980 Long resigned from the Massachusetts General Hospital. Long's studies were supported by a three-year grant from NIH worth $200,000 and which was renewed in 1979 by an additional $500,000. When Long admitted falsifying data, the grant was terminated.

The matter did not end there. Another shock relating to Long's work on Hodgkin's cells came when Nancy Harris and her colleagues published a paper in *Nature* in January 1981[202] in which they demonstrated that, of the four lines established by Long as Hodgkin's lymphoma cells, three were actually identical with a lymphoid cell line from an owl monkey, whereas the fourth Long's line was identifiable only as a human cell distinct from HeLa cells. (HeLa cells were originally obtained in 1951 from a cervical tumour of one Helen Lack, and were then propagated in practically every tissue culture laboratory in the world.) Its identity as a Hodgkin's cell line, however, was doubted, since the spleen from which the cells were originally isolated was found to have had no tumours.[203]

The suspicion regarding the identity of Long's cell lines had originated even before his resignation. Early in 1979 Long sent some samples of these cells for karyotyping (determination of the number and the shape of chromosomes in cells) to Stephen O'Brien at the National Cancer Institute. O'Brien informed Long that all three cell lines were of the same origin, and not different as Long had supposed, and that their enzymatic patterns did not fit those of the red blood cells of the patients from whom the original Hodgkin's cells were taken. When Long applied to the NCI that year for a renewal of his grant, he mentioned in the application that his cells had an HeLa marker in them; this remark was overlooked by the peer review panel. The panel should have suspected that Long wrongly identified the cells as Hodgkin's cell, and should have at least demanded clarification from Long. Long, however, failed to inform his collaborators about the possible contamination of the cells by some other cell line, and therefore nothing was done about O'Brien's report until after Long's resignation. The cells were then sent for karyotyping to several laboratories (University of California School of Public Health and Child Research Center in Michigan) where it was found that Long's cells were those of an owl monkey. Retrospectively, it was discovered that at the time Long isolated the cells from his Hodgkin's patients, there was work going on in the laboratory with the owl–monkey cells.

Like Felig, Long appeared before the subcommittee on Investigations and Oversight of the US House of Representatives, and there he stated:

> In my own instance of wrongdoing the invalid and incorrect findings that I published were brought to light because of the insistence of my colleagues that dishonesty of this sort is not to be tolerated . . . The loss of my ability to be an objective scientist capable of working critically and honestly in the laboratory cannot, in my case, be linked to defects in the system under which I worked. Rather it seems to me that the system has worked effectively to correct the misdeeds of an errant investigator.

Answering a question about the peer review system, Long said that it was impossible for a reviewer to evaluate experiments that had not been done. It was only after the paper was published that corrective action could be taken. (This statement is not entirely correct. There have been many cases when, as a result of suspicion of misconduct, manuscripts have been aborted before publication.)

Dr Lamont Havers, the Director of Research of the Massachusetts General Hospital, testifying before the same committee said:

> Even if the results of Long had been confirmed by others, and thus his conclusions valid, there remains the ethical question that the data upon which he reached these conclusions had been manipulated.

Analysis of the affair makes it clear that Long was guilty of falsification of an experiment in May 1978. As to the identity of the cell lines, the criticism of Long has to be mollified. There might have been an innocent deception, based on an error of judgement and perhaps negligence in following the matter to the end. Had it not been for the discovery of Long's misdeed by his colleague Quay, the whole story of the identity of the cells Long worked with would have been tackled in an entirely different manner, and attributed only to error. After all, anyone working with cell lines knows well the difficulties in making absolutely certain that the line has not been contaminated.

What surprises me in reviewing Long's story is that in all the press reports about Long's misdeed there is no mention of those collaborators who had put their names on the false paper. One would think that if someone puts his name on a paper he must be regarded as a partner in any misdeed. The censure by the press should have taken this into account, but it had not. As to the co-authors of the *Journal of National Cancer Institute* paper all, including Long, have indeed sent a retraction to the editor.

After his resignation from the Massachusetts General Hospital, Long went back to medical practice. One would like to think that scientists in basic research (as distinct from clinical research), devoid of professional contact with the public at large (for instance, patients), would be less prone than their clinical brethren to the temptation of cutting corners and obtaining quick results at the expense of their ethical principles and integrity.

The attribute of being human, and therefore being driven by ambition, desire for power, fame or money, seems in some cases to overrule the ethos of science. Thus, not only in clinical but also in basic sciences we encounter contrived experiments and falsified results, as the following cases will illustrate.

7

Documented Cheating in
Basic Research

THE CASCADE FIASCO

At the May 1981 meeting of the Tumor Virus Group at Cold Spring Harbor, on Long Island (NY) a young graduate student, Mark Spector, offered an exciting and provocative scheme elucidating the sequence of events that connected the infection of animal cells by certain viruses with the transformation of these cells into tumour cells. It seemed to provide a plausible account of how viruses, long studied as possible causes of cancer in humans and other animals, triggered off the change of normal cells to malignant ones.

The implications of this spectacular scheme – called the 'cascade hypothesis' – demand some background understanding of cellular biochemistry. The hypothesis centred on four new enzymes functioning in sequence as protein kinases (enzymes attaching phosphate groups to specific amino acids on proteins). These enzymes, PK_F, PK_L, PK_S and PK_M, usually inactive, were supposed to be linked in such a way that the activation of any one of them would affect in turn the efficiency of a key membranal enzyme, sodium, potassium dependent adenosine triphosphatase (Na + K + –ATPase). This enzyme, deriving energy by removing one phosphate group from adenosine triphosphate (ATP) and forming adenosine diphosphate (ADP), pumps sodium out of the cells so as to keep its concentration in the cells constant. Spector's theory was that in normal cells these enzymes are present, but inactive. When activated, the enzymes phosphorylate tyrosine residues in each other down the line. This sequential activation eventually reaches the ATPase, which, in distinction from the four cascade enzymes, loses its activity when phosphorylated. A decrease of ATPase activity has in fact been observed in tumour cells but not in normal cells. When the ATPase becomes less efficient in tumour cells, they have to use more energy (ATP) in order to provide the required level of sodium pumping. More ATP is thus broken down into ADP and free phosphate. The excess ADP is now used by the cells for 'superfluous' fermentation of glucose to lactic acid, as is indeed observed in tumour cells. This anaerobic conversion of glucose to lactic acid has been called the 'Warburg effect'.

The importance of phosphorylation, especially via tyrosine (rather than via serine) has already been observed during the transformation of normal to tumour cells by certain RNA tumour viruses. These viruses are known to contain protein kinases; for example, the Rous sarcoma virus contains a protein kinase (pp60^{src}) which is a product of one of its genes, the src gene, which is responsible for changing normal cells to tumour ones.

In his lecture at Cold Spring Harbor (as well as in the publications that followed)[204, 205] Spector showed that the protein pp60^{src} produced in cells infected by Rous sarcoma virus had properties identical to the newly discovered cascade enzyme PK$_F$, and thus linked the virus infection to the cascade sequence, to the observed loss of efficiency of ATPase and thence to transformation.

How had this elegant sequence of events been elucidated?

Mark Spector, a graduate student from Cincinnati, had been accepted by Professor Ephraim Racker at the department of Biochemistry and Molecular and Cell Biology at Cornell University. He arrived at Cornell with excellent recommendations, and was put to the task of purifying the membranal ATPase. It was already known then that the ATPase in tumour cells had been less efficient than that in normal cells, but the evidence was circumstantial and was based on the degree of inhibition of the enzyme by various inhibitors (such as ouabain, dinitrophenol etc.), on the movement of radioactive rubidium across the cell membrane, and on lactate production by the cell. There was a need for direct demonstration of the properties of membranal ATPase derived from tumour cells. This was what Spector achieved. Within two months of his arrival at Cornell, he had isolated and purified the membranal ATPase from Ehrlich ascites tumour cells. He then reconstituted the enzyme into liposomes (artificial cell membranes) and showed that the sodium pumping activity was less efficient in liposomes equipped with ATPase derived from tumour cells. In the next step he demonstrated that the decreased efficiency of the tumour ATPase was due to the phosphorylation of the beta subunit of this enzyme; the removal of the phosphate from the enzyme restored its activity.

Spector then looked for an enzyme which phosphorylated the ATPase. He isolated the kinase PL_M, which was indeed able to phosphorylate normal ATPase and decrease its efficiency. Next Spector found that the new enzyme PK_M would phosphorylate ATPase only when it was itself phosphorylated, and this led to the isolation of the next kinase, PK_S, which also required to be phosphorylated for its activity. In this fashion, within six months Spector had isolated four kinases.[204] This feat was considered most unusual for a graduate student, but was ascribed to Spector's being one of the most brilliant young scientists of the day, very hard working and completely devoted to science.

The cascade mechanism, as uncovered by Spector, was interesting *per se* but it became even more important when Spector demonstrated that one of the four enzymes, the PK_F kinase, was identical to the product of the transforming gene from the Rous sarcoma virus ($pp60^{src}$). This protein had been discovered some few years earlier by Ericson at the University of Wisconsin. Spector and his co-workers[205] showed that $pp60^{src}$ and the PK_F kinase had the same molecular weight, were both phosphorylated via tyrosine, were precipitated by the same antiserum, were able to bind to PK_L and to vinculin and to phosphorylate these compounds, and that both acted as a substrate for PK_S (Figure 7).

There was a problem related to the finding that all four enzymes were present not only in tumour cells, but also in normal brain cells, and yet in the brain cells the ATPase was not phosphorylated. Spector attributed the lack of activity of these enzymes in normal cells to the presence of a small polypeptide.[205]

On the basis of all these findings, Spector proposed a hypothesis: when a virus infects a cell and causes it to produce the viral src protein, this protein, being a kinase, trips off the cell's otherwise inactive cascade of kinases, which in the end phosphorylate the membranal ATPase.

Spector's findings raised enormous interest among tumour biologists and virologists, and it was obvious that many would try to repeat his experiments. Among them was the discoverer of the $pp60^{src}$ kinase, Ray Ericson. He spent Spector some antisera he had prepared against his own src kinase to be tested against Spector's PK_F, and in exchange he received from Spector his antisera against PK_F to be tried against $pp60^{src}$. If Spector's scheme were right, then

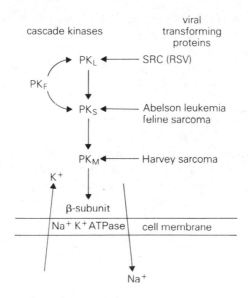

Figure 7. Spector's cascade hypothesis.

his antiserum to PK_F should have precipitated only the src protein. Ericson found, however, that Spector's serum was also active against whole virus particles of Rous sarcoma, and this raised some suspicions.

More disturbing evidence came from the laboratory of Dr Volker Vogt at Cornell. Vogt was the co-author of a paper published in *Cell* with Spector, Racker and Vogt's graduate student, Pepinsky. It was Pepinsky who actually collaborated with Spector in the laboratory. Vogt was worried because he could not repeat some immunoprecipitation experiments with the ATPase in the membranes, although the same experiments performed in Spector's laboratory in the presence of Pepinsky gave the expected results. Again, though Spector demonstrated that the src protein was identical with the PK_F, the immunoprecipitation experiments came out negative. Only when Spector himself was running the gels (on which the proteins were separated and then immunoprecipitated), did the results come out as predicted.

At the end of July 1981 Vogt managed to get hold of one of Spector's gels. The phosphorylated proteins had to be located in the gel by the presence of radioactive phosphorus ^{32}P supplied in the phosphorylation reaction, in which radioactive ATP donated the radioactive phosphorus atom to the protein to be phosphorylated. The product of the phosphorylation would then migrate in the gel to an identifiable spot (Figure 8). When such a gel with ^{32}P labelled protein is laid on a photographic film, the radioactive spots (bands) show up as darkened areas.

To his dismay, Vogt found that in Spector's gels, the radioautographic pattern indicated that the radioactivity of the protein bands was that of iodine, and not of ^{32}P.

It is known that one can tag proteins with radioactive iodine (a process called conjugation) by a simple chemical reaction which does not involve any enzyme. On the other hand, phosphorylation is a reaction mediated by an enzyme, a kinase. The presence of ^{125}I in-

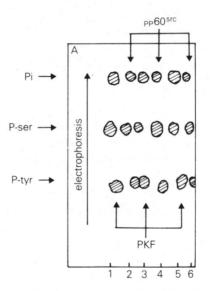

Figure 8. Spector's cascade experiment to show that the enzyme PK_F and the oncogenic protein pp60^{src} were identical, and that both were phosphorylated in their tyrosine residues. Panel A (redrawn from Spector, M. *et al.* 1981: *Cell*, 25:9, figure 8) represents autoradiographs of gels in which the respective proteins were electrophorized after hydrolytic digestion. In panel A, lanes 1, 3 and 5 contained PK_F, while lanes 2, 4 and 6 were pp60^{src} isolated from ^{32}P-labelled cells. PK_F was obtained from uninfected Ehrlich ascites cells, and pp60^{src} from Rous sarcoma virus transformed cell lines. The black spots show the presumed phosphorylated amino acids (serine and tyrosine). Vogt showed that the blackening of the autoradiographed gels was due to the radioactivity of iodine and not phosphorus.

stead of ^{32}P in the gel was therefore an indication of manipulation to produce the desired results.

Vogt contacted Racker and they confronted Spector with the findings: although Spector did not deny that there was iodine in the gel instead of phosphorus, he did not offer any credible explanation for this impossible metamorphosis. Racker asked Spector to prepare new batches of the cascade enzymes and he personally supervised the experiments. He confirmed that the final step of the cascade was correct and that the small polypeptide which activated some kinases indeed existed. Other results claimed by Spector, however, could not be reproduced and thus the question of which of Spector's experiments were faked and which were not was not resolved.[206]

On 9 September 1981 Spector withdrew his PhD thesis and left Cornell. Vogt, Pepinsky and Racker, who had been the co-authors of the paper in *Cell*, retracted their findings in a statement in the same journal:

> This . . . makes us doubt the correctness of the results published in CELL in July 1981 . . . we believe that immunological identification of PK_F and phosphoaminoacid analysis may not be right.[207]

After the forgery had been exposed, Spector's credentials were checked. It was found that, as a student in Cincinnati, he had been sentenced to a suspended prison term for forging his employer's signature on two cheques made out to himself. Another investigation revealed that a paper Spector had published with C. Douglas Winget[208] was also under suspicion of having been fabricated by Spector. In this paper the authors claimed the isolation and purification of an enzyme from chloroplasts responsible for the photosynthetic splitting of water to release oxygen. When the wrong doing of Spector became evident, Winget tried in vain to replicate this exciting experiment. This was thus another footprint in Spector's falsification trail.[209, 210]

In the Spector case, as in the Summerlin and Darsee affairs, we again encounter a situation where a hard-working young scientist, operating in a prestigious laboratory headed by a well-known and respected senior scientist, has the opportunity, and uses it, to falsify the results of experiments and achieve high credibility because of the co-signature of the senior partner.

What is ironic in the Spector case is the fact that some 20 years earlier Professor Racker himself had exposed a scientific fraud committed under similar circumstances by George Webster in the laboratory of Dr David Green in Wisconsin. Webster had claimed[211,212] that in the course of his experiments on oxidative phosphorylation in mitochondria, he had isolated high energy intermediate compounds ($CoQH_4$ of cytochrome C, and coupling factors): these factors, to-day called kinases according to Webster, phosphorylated ADP to ATP, each at one of its three phosphorylating sites, but they did not have ATPase activity.

In Racker's own studies at that time the coupling protein did have ATPase activity. In order to compare and verify Webster's findings, Racker went to Green's laboratory and there found that Webster could not produce any notebooks with experimental details about his compounds. When confronted by Racker, Webster admitted fabrication and retracted his claims. He subsequently abandoned an academic career and became a librarian.

It is indeed a tragic and at the same time ironic fate that the very scientist who so sceptically approached the alleged findings of Webster concerning an enzyme involved in the metabolism of ATP later himself fell prey to a similar confidence trick played by a young man who had miraculously produced at will all the enzymes required by Racker's hypothesis.[213-215]

Spector's fraud was exposed within 16 months of the presentation of the cascade hypothesis to the scientific community at Cold Spring Harbor. During the initial period, after May 1981, various interested research groups first tried to repeat Spector's complicated experiments. When they did not succeed, they tried to find the reasons for the lack of their success. All this took time. The fact that the exposure came within a year or so of the presentation of the false data indicates that the self-policing system of science was working quite well and after a brief fraudulent disturbance, the record was set straight. In fact, in B. M. Sefton's review published in *Current Topics in Microbiology and Immunology* 1986 (123: p.40) on viral protein kinases involved in transformation, Spector and Racker are not mentioned at all.

AN INVESTIGATION THAT TOOK TEN YEARS

Unlike the previous story of Spector's quickly discovered misdeed, a case of unethical behaviour in a university setting concerning research on transplantation of kidneys was resolved only after protracted investigations. It involved Dr Zoltan J. Lucas, Associate

Professor of Surgery at Stanford University, a doctor/scientist with a degree in biochemistry from MIT and MD from Johns Hopkins University.

In 1970 Lucas was working on immunological aspects of kidney transplantation in rats. His research was based on an interesting phenomenon of immunological enhancement: kidney grafts transplanted from one breed of rats to another survived better when the recipient animals were treated at the time of grafting with blood serum of rats immunized against white blood cells of the recipient.[216]

In September 1970 Lucas presented a report on this research at a meeting of the Transplantation Society in the Hague.[217] The first author of this report was a medical student, Randall Morris, working for Lucas at Stanford. Morris also attended the Hague meeting, but he was surprised to learn from Lucas's presentation of successful experiments which he knew he had not carried out prior to his leaving on a two-week vacation before the meeting. When Morris questioned Lucas, he explained that he had performed the reported experiments himself, during Morris's vacation, with the help of a post-doctoral fellow, K. Enomoto. Morris wondered about the speed with which Lucas had done the experiments, but continued to work in Lucas's laboratory throughout 1971.

It was then that he came across an abstract being prepared by Lucas for a meeting of university surgeons. Experiments for which Morris was nominally responsible were being reported in the abstract as completed, with certain results, when Morris knew they had not yet been carried out. After an argument with Lucas over the matter, Morris left Lucas's laboratory and continued his research at the Cardiac Transplantation Laboratory at Stanford, headed by Dr Eugene Dong.

News of the Morris–Lucas disagreement reached the chairman of the Department of Surgery via Dong, and as a result a departmental committee was appointed, headed by Dr Sidney Raffel, to investigate the allegations against Dr Lucas.

A year went by during the course of which a Dr Baronio Martins, then employed by Lucas on an NIH grant, was discharged, and filed a grievance with the university. Before the formal hearing, Martins obtained a copy of the grant progress report and found therein statements about the quality of his work which he considered derogatory. He also noticed that the report cited experiments where the number of samples supposedly used was twice what Martins knew to be correct.

Meanwhile, the ad hoc departmental committee reported that the allegations of Morris and Dong against Lucas were not substained, and that no further action would be taken by the Department of Surgery. Martins, however, filed another complaint with the chairman of the Department of Surgery in February 1975, and a new committee was formed under the chairmanship of Professor George Feigen (members Prof. L. A. Herzenberg and Prof. R. Mishell). The committee's report (2 October 1975) stated:

> . . . the committee formed the distinct impression that Dr Lucas' laboratory was incompetently run, the personnel undisciplined, and the note taking and record keeping either non-existent, or chaotic to such a degree as to make it impossible to verify certain crucial points in connection with the work alleged to have been done by Dr Enomoto or others.[216, 218] The committee believes that as much of the impropriety resulted from scientific and administrative incompetence as from deliberate attempts to fabricate.

That summer Martins filed a suit in the superior court of Santa Clara County (No. 372325) claiming libel and bad faith by Lucas and the university authorities. He alleged that he had been fired by Lucas because his experiments did not produce results identical with Lucas's. The case dragged on until January 1981, when it was settled in favour of Martins.[219] What had gone on between 1975 and the 1981 settlement?

In December 1975, Dr Eugene Dong, whose complaint in support of Morris had led to the formation of the Raffel and Feigen committees, wrote to the US Assistant Secretary of Health, Dr Theodore Cooper, pointing out the discrepancies in the Enomoto kidney transplant reports. The reply came from the Deputy Director of NIH, Dr Ronald Lamont-Havers, who concluded that '. . . there appears to be no pretence or deception or attempt at withholding relevant data from NIH reviewers'. This did not satisfy Dong, who wrote another letter to NIH in March 1976. This time he was informed that there was a special committee at Stanford investigating the allegations against Lucas, and NIH 'does not feel it should impose itself into that investigation'.

In fact, there was an extensive exchange of materials and letters between Stanford and NIH, and finally Dr Leon Jacobs, Associate Director of NIH, wrote: 'I believe the matter should be closed. If Stanford informs NIH that there is evidence of false research reporting to NIH then we can re-open the matter directly with Stanford.'

When, in October 1976, NIH requested a progress report, the university replied: 'The committee was convened and has submitted a report to James B. D. Mark, MD, acting chairman of the Department of Surgery. Though the distribution of the report was promised for January 1977, Dr Ralph Currens of NIH had to inquire again in May as to the progress of the review. In June, Stanford informed NIH that the Morris allegations were not substantiated and that the Feigen committee had issued a report which concluded that 'Dr Lucas was not guilty of fallacious research but perhaps was derelict in not maintaining good laboratory administration' (letter from J. J. Schwartz, Stanford University Council to J.W. Schriver, NIH, of 25 January 1979); this statement was made in spite of committee's conclusions (as quoted on p. 106) and remarks such as 'the committee believes that as much of the impropriety resulted from scientific and administrative incompetence as from deliberate attempts to fabricate'. In fact, the committee suggested 'continuation of the enquiry by an appropriate agency of the University'.

In the meantime, in January 1977, in the course of the court case *Martins* v. *Lucas*, Lucas had been requested to provide manuscripts which, according to his grant applications, had been submitted or accepted for publication. Thus it was discovered that some of these manuscripts cited as having been published either did not exist, or had been rejected by the journals quoted. The laboratory notebooks which had also been provided revealed that the number of animals used per experiment was only half that stated in the grant proposal.

Dong used this information in a letter he wrote in March 1977 to Dr Lawrence Horowitz, Chief of Staff of the Senate Subcommittee Overseeing NIH, a committee headed by Senator Edward Kennedy. The letter called Dr Horowitz's attention to numerous data discrepancies in the paper Lucas published in 1974 in the *Journal of Immunology* with his PhD student Sharyn Walker.[220] Senator Kennedy requested clarification from Stanford attorneys in June. When, later, Walker was subpoenaed for the notebooks relating to the published paper, she stated that they had been lost when she moved to a new house.

In August, on the basis of the accumulated information, Dr Ralph Currens of NIH travelled to interview Dr Walker. She told Currens that Lucas was a dedicated and exceptionally brilliant man. Dr Currens reported that Walker 'felt that Lucas anticipated results in place of waiting for the final result, and as a consequence probably anticipated writing these papers, but as progress in their work went on, he changed his mind and combined the work in other manuscripts'.[221]

Lucas himself, in a letter to the President of Stanford (7 July, 1979) wrote:

> As I indicated in my previous letter, I realize that the semantics with which I expressed the state of our work were erroneous. However, they resulted from being passionately caught up in the work they represented. It was a recurring, non-random event. In each of the reports, when I was involved in work or the preparation of the work for publication, I was so sure that they would be completed and submitted in the next several weeks that they were reported as such.

Max[222] remarked that there are 'always a number of small steps between the completely unacceptable and unethical behavior'. One such step, described in Lucas's letter, comes in supplying an intelligent guess to a result of an experiment in the hope that by the time the paper is presented at a conference, the results will have come out as expected.

In September 1977 the Dean of the Stanford Medical School, Clayton Rich, informed Dr Donald Frederickson, the director of NIH, that some of the Lucas bibliographic references were incorrect, and he in turn reported to Senator Kennedy that 'there was an overrepresentation of accomplishment' by Lucas, but it was not determined that there was a deliberate misrepresentation or falsification.

In June 1978 Lucas applied for a 6-month sabbatical leave, and at the same time submitted a grant/contract application under a cover letter from Lawrence Crawley, the Vice-Dean of the Medical School at Stanford.

Lucas's notebooks, subpoenaed in the Martins–Lucas case in 1978, were examined in 1980 by Dr Dong and a forensic specialist, Dr Kenneth Parker. Visual inspection, as well as special laboratory techniques, revealed that while for most entries in the notebook a ballpoint pen was used, the 'fabricated' entries might have been made with other types of pen (*Peninsula Time Tribune*, 19 January 1981). Suspicions were aroused on finding supposedly old and recent entries made with the same ink. The deaths of rats in the experiment records were marked with an X; in some cases, however, the X had been erased or cut away with a razor, and continuous straight lines, indicating surviving animals, had been drawn across. Rat survival records were indicated by straight broken lines, increased after each day of observation. For the allegedly falsified rats, the lines were found to be continuous all through the experimental period. Irregularities were also found in the analyses of the serum of the rats (*San Jose Mercury*, 6 September 1981). Parker concluded that as many as 30 per cent of the rats listed did not exist, and the lives of other animals had been either shortened or prolonged with the stroke of a pen.

In March 1979 Stanford University informed NIH that during the preceding ten years Dr Lucas had submitted 30 false citations of his work. Lucas himself admitted only that 18 of these papers, published between 1967 and 1977, were in error: ten were rejected, five never submitted, and in three cases the wrong journal was quoted.

The university reprimanded Lucas and suspended him without pay for 12 weeks (*San Francisco Chronicle*, 7 March 1979; *Campus Reports*, 7 March 1979). University President Richard W. Lyman said on this occasion: 'There has been no conclusion that your research results have been seriously affected. The trouble with carelessness in research, however, is that one has no way of knowing when it may have devastating consequences' (Stanford University News Service, 3 June 1979).

In spite of the investigation Lucas received awards of nearly half a million dollars: he was a principal investigator in two NIH projects, one expiring in 1980 and the other in 1981. The first grant from NCI was given for a year in 1978 and then extended till September 1980. The second grant began in January 1979. In October 1981 Lucas took leave of absence and eventually left Stanford.

The question that arises for us here is why did it take six years and three committees before a university professor was cited for misconduct? Vice-President for Public Affairs at Stanford, Robert Rosenzweig, said to the *San Francisco Chronicle*: 'There are situations in which the true answer does not sound sensible and the true answer is that it was sloppiness, inattention, a creaking of wheels.'

The lesson to be learned here is that all institutions should have efficient machinery for the prompt investigation of suspected misdemeanour (see chapter 15). If the charges are unfounded, speedy action will stop the proliferation of rumours likely to harm the innocent. Quick detection of misconduct may well have a deterrent value.

Many of the cases of fraud we have been discussing, in this and the preceeding chapter, concern junior scientists, such as doctoral students, or scientists completing their training as researchers by gaining practical experience in a laboratory of a prominent scientist. In cases where the guilty party is a junior researcher publishing with his senior, the take-home lesson seems clear: the research supervisor has to supervise. A brisk morning visit to the laboratory to inspect the latest results is not good enough; on the contrary, it puts dangerous pressure on the juniors, which some may be unable to bear. The fact that the name of the supervisor appeared, as we have seen, on a number of papers based on hypothetical data from experiments that had never been performed should unequivocally inculpate the supervisor no less than the junior. Senior authors should be in a position to stand by the data presented. Junior authors should be obliged to keep detailed and permanent records of all experiments, including all raw data.

The situation is more complex, however, when the misconduct is on the part of an acknowledged scientific investigator. What motivates such a senior scientist in his transgression? To a certain degree the blame can be laid on the contemporary scientific community, with its escalating demand for quantity in research. Lucas might not have falsified his publication list had he not apparently been confident that what counted in the NIH refereeing system was quantity, and considered it unlikely that papers would be checked for veracity and quality.

WAS T-RNA CRYSTALLIZED?

Another example of scientific misconduct emanating from a senior scientist is the case of Professor Hasko Paradies. He came to prominence in 1967, when he described the production of crystals of valine transfer RNA (t-RNA) from yeast. Transfer RNAs are molecules that carry amino acids (the 'building blocks' of proteins) to messenger RNA (m-RNA) being processed on the ribosomes. There is one type of transfer RNA for each of the 20 amino acids. A code (a triplet of nucleotides) on the messenger RNA attached to a ribosome instructs the t-RNA which amino acid to bind onto the growing chain of a polypeptide.

At the time Paradies was working at the University of Uppsala in Sweden.[223] In 1970 he conducted further research at King's College in London and from there published the first diffraction patterns of t-RNA crystals.[224] (The arrangement of atoms in a crystal can be deduced from diffraction patterns produced by X-rays passing through the crystal. These rays are bent and reflected by the atoms in the crystal and their image is projected on to photographic plates. The regular and symmetric but characteristic distribution of black spots on

Charles Babbage

Alexander Gurwitch

Ptolemy

Sir Isaac Newton

Gregor Johann Mendel   George O. Gey

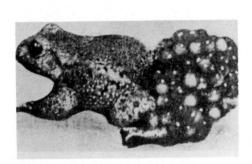

Alytes obstetricans, the midwife toad

Paul Kammerer

Alexis Carrel

Sir Cyril Burt

Trofim D. Lysenko

Harvey Fletcher

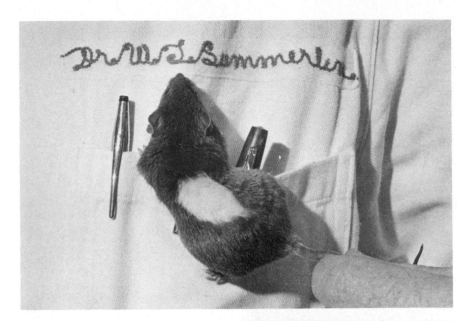

Above: 'Old Man', a black mouse with successfully transplanted skin from a white mouse donor. Summerlin failed to notice that the black mouse was not pure bred but a hybrid *Right:* William T. Summerlin

Above left: Robert Good   *Above right:* Marc Spector   *Left:* Philip Felig   *Below.* Margaret Mead in Samoa

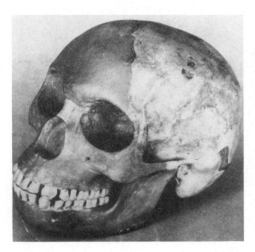

The Piltdown skull (*Eoanthropus dawsoni*) reconstructed from the cranium and jaw found by Dawson, exhibited at the British Museum

Forged figurines, presumably made by Salem, an accomplice of Shapira. Similar figurines with Moabite script were sold to the Berlin Museum

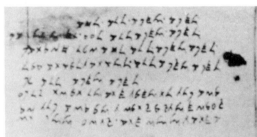

Above: Parchment scrolls, as exhibited at the British Museum when Shapira offered them for sale. Clairemont-Ganneau proved them to be forgeries.
Right: Etruscan sarcophagus

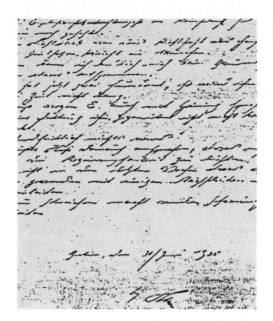

Above: Alsabti's signature from a greeting card   *Left:* Hitler's forged script   *Centre:* Titles and abstracts of Ferguson papers in *Science* and *Experientia*   *Bottom:* Lock's remarks in the *BMJ*

Extrinsic Microbial Degradation of the Alligator Eggshell

Abstract. *The outer, densely calcified layer of the alligator eggshell shows progressive crystal dissolution, with the production of concentrically stepped erosion craters, as incubation progresses. This dissolution is caused by the acidic metabolic by-products of nest bacteria. Extrinsic degradation serves to gradually increase the porosity and decrease the strength of the eggshell.*

"Notice of inadvertent repetitive publication: The *BMJ* regrets that the article entitled 'Use of lasers in pinealology' by AC Block and YZ Tackle of the Medical College of the University of the Scillies (30 February 1983, p 1937) was substantially the same as 'Pinealology and laser use' by YZ Tackle and AC Block published in the *British Journal of Pinealology* (1983; **18**: 122–8). The authors hold sole responsibility for this action, which is in violation of accepted scientific ethics and of the *BMJ*'s Instructions to Authors.''

TRICKS OF THE TESTING TRADE

Private testing laboratories live and die by the *positive* results generated on products they test for safety. Your future as a laboratory owner or administrator is in serious jeopardy if you find too many products unsafe. But fortunately, manufacturers rarely run their own audits on tests contracted out to private labs. That way, they can claim ignorance of wrongdoing if protocol is not followed. So here are some tricks you can use to help that drug, chemical or product meet approval standards.

THE HAPPY ENDING. Bureaucrats rarely seem to have time to read more than summaries and conclusions. So you can do your manufacturing client a great service by having the conclusion bear minimal relationship to your findings, particularly if the findings are negative. Remember that statistics are at the service of those that use them.

THE SHRED. Keep a paper shredder near your records at all times. If you hear that a government inspector or auditor is about to visit, use the shredder quickly to rid your files of embarrassing or incriminating data. For fast shredding techniques, call Industrial Bio-Test Laboratories.

THE BENCHBOOK BOOGIE. If you have to create new benchbooks to substantiate your hypothesis, remember to use several different pens and pencils to make it look as if figures were entered over the duration of the study, not during one frantic night at home on the kitchen table. If your benchbooks are recording data from animal experiments, a dab or two of rat feces along with a splash or smudge of blood adds to the authenticity of the book. Dr. Ron Smith could give you more hints on creating raw data from whole cloth, but your guess is as good as ours as to his whereabouts. (See box on page 49 for more on Smith's methods.)

THE INCOMPETENT STAFF. There are absolutely no standard professional or licensing qualifications for research lab technicians who do safety testing. So aside from the obvious cost savings gained by hiring untrained technicians, you can always use them as an excuse when the scientific method is inadvertently abused or abandoned. You can contact almost any testing lab in the country in your search for undereducated, unqualified and incompetent staff. Move your staff around the lab constantly throughout experiments. If someone is leaving your firm, have *that* person sign all the benchbooks. That way the "responsible party" will be absent should an inspection ensue.

> **K**eep a scalpel in your lab, a shredder near your files, and don't tell the Feds your data caught fire.

THE SPEEDUP (OR THE TIME-LINE FUDGE). Many safety tests require long exposure of the product to hostile environments like extreme heat, dryness, cold or wetness. Take semiconductors, for example. Baking different batches of semiconductors for periods from three hours up to ten days is time-consuming and expensive. Why not just bake them all for three hours and record them at different durations? For complete instructions on time-line fudging, contact the National Semiconductor Company in Santa Clara, California.

THE SCALPEL. There's nothing like the intervention of modern microsurgical techniques to "cure" laboratory animals that develop tumors from the chemicals they ingest. Tumors can be excised from live animals and never reported in test summaries submitted to the FDA. (The animals are, after all, by the end of the study, alive and healthy.) For instructions on the creative use of the scalpel contact the G. D. Searle Company and ask about the tests it performed on Aldactone, a diuretic prescribed for high blood pressure and fluid retention. Or write to Syntex Corporation and ask about the research it contracted for on the antiarthritic drug Naprosyn.

THE POSTHUMOUS APPEARANCE. Should laboratory animals die, replace them with live, healthy relatives. Again, contact G. D. Searle and Industrial Bio-Test Laboratories in Chicago to learn how some of their rats, monkeys, rabbits and dogs, which died during tests, later appeared alive in test data. Animals can also be recorded as gaining weight and recovering from tumors after their death. Use your benchbook creatively.

THE DATA MASSAGE. Also known as "Graphiting," data massaging is probably the most common fudging technique used in labs today. Motivated by what scientists euphemistically call "intentional bias," it is really the easiest way to get the results you want once the experiment is completed. There is no one way to massage data. Just take those long computer reports home with you, spread them out on the kitchen table and gently apply rubber and graphite.

SOME EXTRA ADVICE: If the validity of your study is questioned, do not under any circumstances use the following excuses. The Feds are familiar with them, and the attempt will arouse immediate suspicion: data destroyed in fire (excuse already used three times); data destroyed in flood (three times); data lost in boating accident, also known as the Andrea Doria Syndrome (once); coinvestigator died (twice); records lost due to hospital closing (three times); records lost due to burglary, robbery or vandalism (five times); data dropped in sewer and destroyed because of stench (once); clinical lab technician dead or missing (once); office nurse dead or missing (three times); nurse or resident did it without my knowledge (14 times).

**BY IRENE MOOSEN
AND MARK DOWIE**

Text from 'Tricks of the Testing Trade'

the photograph can be interpreted in terms of the molecular structure of the crystal.)

In 1974 Paradies joined the Max Planck Institute in Berlin and became Professor of Biology at the Free University of Berlin. From there, in 1977 and in 1979, he published papers describing the crystallization of the carboxylase of a green alga[225] and of the coupling factor CF1 from spinach protoplasts.[226] In mid-April 1983, however, Paradies, now in his early forties, resigned from his chair of biology.

What led to his resignation and the early end of his academic career? It seems that it was the outcome of a university investigation of charges of scientific fraud raised against Paradies by Dr Wayne Hendrickson of the Naval Research Laboratory, where Paradies had spent a sabbatical.[227, 228]

Hendrickson had some doubts about Paradies's work. He carefully inspected the diffraction patterns attributed to valine t-RNA published from Uppsala[224]

> We find that the diffraction pattern attributed to valyl t-RNA in Fig. 2 of that article is in fact an X-ray photograph of a crystal of human carbonic anhydrase (HCAB) . . . we see no alternative but to conclude that this was a deliberate misrepresentation.[227]

Hendrickson assumed that while Paradies was at the Wallenberg Laboratory in Uppsala he had had access to HCAB. Hendrickson does not accept the possibility that the HCAB photographs were erroneously inserted into the 1970 paper since '. . . it seems unacceptably implausible that such a mistake could occur and then remain undetected when Paradies proof-read and subsequently cited the article.'

Newmark, one of the editors of *Nature*, reported that in September 1970 David Blow visited Paradies at King's College.[228] Paradies showed him the diffraction patterns of t-RNA; Blow recognized one of the prints as coming from his own crystallographic study on chymotrypsin. After this incident, Paradies was advised to leave King's College.

Having found the 'misrepresentation' in the t-RNA paper, Hendrickson scrutinized other publications by the same author. He found that Paradies had published diffraction patterns of crystals of seryl t-RNA which were actually those of an enzyme HCAC (human carbonic RNA anydrase).[229] In another paper,[230] he had presented a picture of diffraction pattern of a crystal valyl t-RNA which, in fact was also that of HCAC. In the papers published in 1977 and 1979, the same electron micrograph was used to depict two different crystals, one obtained from algae[225] and one from spinach.[226]

In his reply to the criticism of Hendrickson and others, Paradies wrote:

> The very unfortunate mistake by which the electron micrograph and its optical diffraction were used again in my paper on spinach protoplast coupling factor was corrected as soon as the error was brought to my attention.[231]

Paradies also countered Hendrickson's other charges. He claimed that in his 1970 paper the cell dimension, crystal stability and solubility, and the temperature dependence were consistent with a crystal of t-RNA. He justified the other critized paper by stating that the magnification (which was not indicated in Fig. 2 of the article) was different in the horizontal and vertical planes, an explanation which is hardly credible.

The doubt cast on Paradies' integrity cut short an academic career that had begun so brilliantly 15 years earlier. Nevertheless, one should not forget that Paradies's talent in designing a crystallization protocol permitted Brian F. C. Clarck of the Molecular Biology Laboratory in Cambridge to obtain crystals of t-RNA.[232]

The Paradies case once more indicates that misrepresentation of a 'scientific fact', if sufficiently important, will eventually be discovered, even if it takes a decade or more. I would assume, however, that if the misrepresentation by Paradies were linked to one paper only, it might have escaped scrutiny and detection even after 15 years.

A final point concerning Paradies that should be made clear relates to his answer to his critics.[231] In paragraph four of his reply, Paradies wrote that the optical diffraction was obtained with the facilities of Dr Reuber at the Fritz Haber Institute in Berlin. My own investigations at this Institute revealed that Dr Ellen Reuber had nothing to do with any crystals or pictures from Paradies. He only once asked her on the telephone for some details about the apparatus she was using. Thus, Paradies's statement that the picture of the crystals was obtained via Dr Reuber had no factual basis.

A FIGMENT OF IMAGINATION

A remarkable case of data falsification involved D. Robert J. Gullis. During his career he published 11 papers, seven co-authored by his supervisor at the University of Birmingham (UK), Charles E. Rowe, and four co-authored by scientists at the Max Planck Institute for Biochemistry at Matrinsried, Germany, before he was obliged to admit in *Nature* that his data were a 'mere figment of his imagination'.

Gullis spent two post-doctoral years in the laboratory of Dr B. Hamprecht at the Max Planck Institute. During these years he published, together with other investigators there, papers on the effects of morphine and other neuroactive agents on the levels of cyclic guanosine monophosphate (c-GMP) and cyclic adenosine monophosphate (c-AMP) in cells of tumours of the nervous system. (c-AMP and c-GMP are important molecules essential in the transmission of signals from outside into the cells and in activating in the cells cascades of enzymes necessary for various metabolic processes.)

At the end of his post-doctoral studies, Gullis left the Max Planck Institute. His colleagues there tried to repeat some of his published experiments, but to no avail. Gullis was therefore invited back to the Institute to repeat his experiments. Because of the doubts that had arisen, the samples to be examined in the critical experiments were coded. It was soon found that neither morphine nor the natural sedatives present in brain, the enkephalins, changed the level of c-GMP. Gullis then admitted that his data were fictitious. He also admitted that four papers he had published with Rowe in the period 1973–76 were only hypotheses and were not based on experimental data.

Hamprecht and Gullis then published a joint statement in *Nature* (1977, 265:746) listing the published papers 'which were based on invented data'. Gullis wrote:

> I must take full responsibility for the unfortunate incidents and have consequently suffered. I hope that my experiences are noted by others, and I would like to apologize to the scientific community and the various people involved.

How did it happen that the other members of the team at the Institute (Traber, Moroder and Wuensch) did not suspect Gullis's wrong-doing? Hamprecht's statement in *Nature* said that the print-out data from the scintillation counter were handled only by Gullis. It was he who summed up and evaluated the data based on these printouts. The work of the group was based on mutual trust, and it occurred to none of the other investigators that they should inspect the raw data: as long as the results were sensible and consistent, nobody suspected that the data were being manipulated (*Science News* 1977, 111: 150).

Gullis's career in science was finished. An entire group of research results were wiped off the record, but the printed references to these papers still linger in various publications and reference periodicals, such as *Index Medicus, Biological Abstracts, Chemical Abstracts* etc.

When a misconduct or a forgery is exposed by public admission, as in the Gullis case, all scientists directly involved in the field are informed and thus the record is set straight. Those who may be misled are the newcomers, who enter the field by updating themselves via the abstracting periodicals. When, however, they become seriously involved in the particular area where fraud was perpetrated, they will learn soon enough to discount the fabricated information.

SUGAR STUDIES TURN SOUR

Another confession to falsification of data come from Dr J. Purves of the University of Bristol, who had developed techniques for the study of fetal mammalian brains in utero. At the International Congress of Physiological Sciences in 1981 Purves presented some data on the uptake of 5-deoxyglucose (a non-metabolizable analogue of glucose) by fetal sheep brains.

Animal cells need the sugar glucose as a source of energy and as a building block of more complicated molecules. The cell membranes therefore have a mechanism for transporting glucose from outside into the cell. In order to be able to measure the uptake of sugar by the cells, the scientist replaces glucose by 5-deoxyglucose, a molecule which closely resembles glucose but which is useless when inside the cells because it cannot be metabolized. In Purves's experiments the deoxyglucose was labelled by radioactive atoms and therefore, after its entry into the cell, could be traced and its concentration measured. Purves claimed that deoxyglucose was taken up more slowly in sleeping embryos than in active, awake ones. His work was published in the Congress proceedings.[233]

Purves's junior colleagues raised the doubt that his work was not reproducible. There was an interdepartmental investigation and Purves resigned his post at the university. He also sent a letter to *Nature*[234] stating that the data he had published in the proceedings of the congress were false.

The editors of *Nature* expressed astonishment that a talented scientist such as Purves, whose research was supported by the

Wellcome Foundation and public health services, should have followed a course that led to his resignation and discredit.

A further insight into the motivation for perpetrating a scientific fraud is provided by the story of Dr Joseph Cort, a graduate of Harvard Medical School. He obtained a fellowship for studies in England in 1951, but because of his political affiliations, the US embassy in London wished him to return to the USA for interrogation about alleged subversive activities. He was again recalled in 1953 for induction into the USA army, but refused to comply with these requests. England would not grant him political asylum, so he went to Czechoslovakia where he stayed for 22 years, working as a chemist.

In 1975 the US Government dismissed the indictment against Cort, so he returned to the USA and took up a post as a research and teaching scientist at the Mount Sinai Hospital in New York. While in Czechoslovakia, Cort had worked as an organic chemist producing compounds similar in their structure to vasopressin (a hormone controlling water balance of the kidneys). The modification of the structure of such biologically active molecules may endow them with new useful properties or reduce their toxicity. The analogues synthesized by Cort were DDVAP (prescribed to alcoholics as a drug reducing their ability to consume liquor) and glypressin (for treatment of internal haemorrhages).

At Mount Sinai Hospital, Cort's research involved the design of other vasopressin analogues which would increase the level of Factor VIII in blood (the clotting factor which is missing in the blood of haemophiliacs). Although the production of Factor VIII can be enhanced by vasopressin alone, the latter's side effects, such as its tendency to raise the blood pressure and the decrease the flow of urine, prevented its use on haemophiliacs. Cort's research, supported by Vega Laboratories, led to the synthesis of several vasopressin analogues that would stimulate the production of Factor VIII without the hypertensive and diuretic side effects. Cort patented these new compounds.

In 1980 he gave up his post at Mount Sinai Hospital and moved to Vega Laboratories in Tucson, Arizona, to continue his research. In December 1980, however, he admitted to the president of the company that his Mount Sinai data had been, in fact, a product of his imagination.

Upon this disclosure, a committee of inquiry was formed at Mount Sinai. This committee, made up of two trustees, two outside scientists and six faculty members, found that Cort's reports differed from the data in his laboratory notebooks, and that he had actually done only a fraction of the work he reported.

Cort admitted he did not keep good records. His invention and falsification of data, he said, were dictated by his being

> . . . under a lot of pressure and confusion . . . I had to earn the money for research, or die . . . I knew one could say things in American patent application as long as one said it could be done, and I was close to getting it done anyway. Deliberately, I used the wrong tense. (Quoted by *New York Times*, 27 December 1982, p. B1,B4)

8

Criticism or Slander?

During the past few years certain incidents have been brought to the attention of the scientific community and the general public involving scientists who have been accused of misconduct, be it propagation of a falsehood or forging of their results, but against whom evidence for such misconduct has not been sufficient for condemnation. In such cases, publication of the accusations, be it in scientific journals or in the mass media, before they have been substantiated can do irreparable harm to the reputation of the scientist in question. This is particularly so when accusations are raised against a scientist who is no longer alive and cannot refute inaccurate statements.

Sometimes the borderline between legitimate criticism and slander (or libel) is very thin. The way in which various scientists have been accused in such a manner raises ethical questions that will be discussed in this chapter.

WAS MARGARET MEAD MISLED BY THE SAMOAN GIRLS?

Margaret Mead enjoyed tremendous authority among fellow anthropologists as well as the non-professional public. She published books that have since become classics. Nevertheless, since her death in 1978, voices have been raised by some of her professional colleagues which tend to tarnish her image.

The attacks focus on her famous book *Coming of Age in Samoa*.[235] In this book, which attracted a very wide readership in the Western world as a portrait of a stressless primitive culture, Margaret Mead depicted the graceful life of Samoan adolescents with casual family ties and relaxed sexual mores, a life devoid of guilt. According to her, Samoan children grew up without any strong attachment to parents. Their life was easy and casual, with no conflicts and an absence of adolescent turmoil. Mead's study in Samoa was conducted in the early 1920s, during a nine-month stay on the island shortly after her graduation in anthropology.

Margaret Mead had grown up in an academic environment. Her father was a professor of economics at the University of Pennsylvania and her mother had a degree in sociology. During her studies at Barnard College she was engaged in the feminist movement. Her PhD thesis supervisor was Frank Boas, a German born physicist who switched to anthropology. Mead was influenced by his concepts that not all human traits were genetically determined, but were strongly influenced by nurture. She went to work with adolescent girls in Samoa upon the specific advice of Boas.

During her stay in Samoa, Mead lived with an American family in Manua. She conducted her studies by interviewing some 25 adolescent girls with the help of an interpreter; she obtained most of the material for her book from these interviews. After her return to the USA she obtained a post at the American Museum of Natural History in New York, and from there she published her report on Samoa in 1928.

Mead was the first anthropologist to devote special attention to women and children. She was also a pioneer in the use of photography in anthropological studies. She published some two dozen books and used part of her royalties to support the Institute of Intercultural Studies, which she was instrumental in establishing. She died in 1978, at the age of 76.

A few years after Mead's death, in 1983, Derek Freeman, Professor Emeritus of Anthropology from Canberra claimed in his book[236] that Mead has misrepresented Samoan culture because of 'self-delusion' and 'faulty evidence'; she had, he said, got things 'astronomically wrong'.

Freeman first went to Samoa in 1940 and there studied the Samoans for several years. He learned the language, was actually adopted by a Samoan family and even sat on the council of chiefs. He found that adolescents and children were under considerable stress in an authoritative culture. In clear distinction from Mead's image of Samoans, Freeman found them to be competitive and prone to jealousy and violence. The rate of homicide and rape in Samoa in 1966 was 66 per 100,000, that is twice that in the USA and 20 times that in England. Freeman also stated that Samoan mores prohibited premarital sex, which in Mead's opinion was a casual matter.

Some of Freeman's conclusions about Samoans were further confirmed by Brad Shore, an anthropologist from Emory University in America, who spent five years in Samoa and found that Samoans punished their children by beating them, a behaviour which, according to him, encouraged pent-up aggression. He recognized two contradictory aspects of Samoan society, of which Mead saw one and Freeman the other. Margaret Mead might have been wrong in expecting in a primitive society a degree of homogeneity which we now know does not exist.

Freeman thought that Margaret Mead's view of adolescent free love was based on 'counterfeit tales of casual love' provided by her teenage informants. According to him Mead, as a student of Boas, believed in the doctrine of cultural determinism, namely that most human behaviour is shaped by culture, and very little by heredity. Her erroneous conception of Samoan life was the result of the evidence she assembled to support her ideology; that is, she found what she was looking for. S. L. Washburn, professor emeritus of anthropology, is quoted as saying: "She'd make world shaking, world-wide generalizations from a very small number of people in odd conditions"[237] (also Leo in *Time Magazine*, 14 February 1983, p.52). Freeman claimed that by her dramatization of primitive and happy living in Samoa, Mead had greatly influenced the direction of liberal education in the Western world. 'The tragedy is that Freeman's critique is far more ideological and tendentious than the work he criticizes' (from a letter of Mead's daughter Mary Catherine Bateson).

What is strange about Freeman's disclosures is that, although he long knew the 'truth' and was aware of Mead's error in depicting the Samoan culture, and although he had actually assembled the material for his book in the 1960s, he waited until after Mead's death to publish it. Professor Washburn's view was that it was unfair of Freeman to have published the book only after Mead's death.[237]

In assessing Freeman's accusations, one might remember that between the studies of Mead and those of Freeman some 20–30 years passed, i.e., a whole generation. During this period of international influence by radio, and television, some cultural patterns may have changed throughout the world, as well as in Samoa.

Moreover, Freeman's criticism of Mead is based on a cryptic assumption that standards of ethnography have not changed since 1920. The facts, however, are different. Only a decade after Mead's studies in Samoa, the concept of good ethnographic research already entailed learning the local language, participating in local events,

studying the social system in its actual practice and making a
semantic analysis of the language. It also became good practice to
collect genealogical data about the community, map the land hold-
ings, take censuses etc. One surely cannot blame Mead for not having employed standards that developed *after* her study in Samoa.
The views she had expressed about the relative importance of culture
versus biology can hardly be considered as representative of present-
day anthropology.

Many would feel that Mead's contribution to this science lies in
her being one of the first to question the scientific and lay views
about human behaviour – men versus women, adolescents versus
adults, nature versus nurture. In Mead's day the whole issue of adolescence was based on studies of male Western samples in which
females were occasionally referred to as exceptions. Mead studied
adolescence on the basis of female samples, and this might have led
her to erroneous conclusions concerning the differences between
American and Samoan populations. Methodologically Mead had
done pioneering work, even if the theoretical ideas were flawed.

By stressing the importance of custom and tradition in structuring
behaviour Mead inspired many to a new attitude in research, even if
her own fieldwork would not conform to today's standards. If
Mead's books were not a paradigm of scientific accuracy, they were
widely read and stimulated many students to take up anthropology.

Later criticism has not tarnished Margaret Mead's image and
many would still agree with Goodenough:[238] 'Through the inspiring
role she played Margaret Mead had become a national institution
by the end of her career.' Subjecting her to criticism in a historical
perspective would seem to be unfair.

SWAMINATHAN'S SUPERWHEAT

With the unmasking of some well-publicized cases of scientific fraud
has come a climate of suspicion with regard to the activities of scientists, particularly so in biology and medicine. Accusations have
appeared in the mass media, and even is some scientific journals,
directed against a number of well-known scientists before it has been
established whether they were in fact guilty of any misconduct.
Among scientists so accused is Dr Monkombu S. Swaminathan.

In the 1960s and early 1970s Swaminathan was the director
general of the Indian Council of Agricultural Research, and a

member of a consultative group on agricultural research of the Food and Agriculture Organization (FAO). Swaminathan had received in 1963 several strains of dwarf wheat, among them one, Sonora 64, which was considered to be very promising. The strain was unacceptable in India, however, because its red colour made it unsuitable for baking bread. Swaminathan's team irradiated Sonora 64 seeds with gamma and ultraviolet radiation to produce mutants. One such mutant, obtained in 1967, was Sharbati–Sonora, which had the proper amber colour and better baking properties.

In November 1974 Dr J. Hanlon published an article in the *New Scientist* entitled 'Top scientist published false data'.[239] In this article, based on information obtained from a disgruntled former colleague of Swaminathan, Hanlon attacked Swaminathan's claim that the Sharbati–Sonora had a 16.5 per cent content of protein and an unusually high lysine content (4.16 per cent). Repeated analysis of the amino acids, however, carried out by scientists at Purdue University (Lafayette, Indiana) at the end of 1967, showed that the actual lysine content of the Sharbati–Sonora wheat was in the same range as that of the parent strain, Sonora 64, namely 2.21–2.83 per cent. Hanlon reacted sarcastically to Swaminathan's original announcement that the new strain had a significantly higher lysine content.

Referring to the induction of mutations by ionizing radiation he wrote:

> . . . this announcement showed that third world scientists could combine the best products of the scientific revolution with the most sophisticated nuclear technology and produce something even their experts thought impossible.

Hanlon concluded his article with the following question:

> Should Swaminathan be excused because he is an able scientist who outstretched himself and published inadequately supervised work done by graduate students, as Riley argues, or is it, as Silow sees it, dangerous to have sitting on the highest level advisory bodies a scientist who has so extensively published so much non valid science in those very fields?

J. Hutchinson[240] reacted to this statement in the columns of *New Scientist* saying that it was 'monstrous' to pose such a question without waiting for a reply to the allegations. Dr Austin, principal investigator of wheat quality at the Indian Agricultural Research Institute in Delhi wrote a letter to the *New Scientist* terming Hanlon's article 'an attempt to slander rather than at scientific criticism and analysis'. The reported 4.61 per cent lysine content, said Dr Austin, was due to an experimental error caused by the wrong acidity of the buffer used in the analysis based on decarboxylation of the protein hydrolysate. This erroneous result had been mentioned by Swaminathan in a lecture at a Vegetarian Congress in Delhi in 1967 and in a paper in *Food Industries Journal* published in November the same year. In three later publications, however, the value had been corrected to a range of 2.57–3.19 per cent. It was regrettable indeed, wrote Austin, that 'Dr Hanlon has chosen to malign our extensive work on quality improvements of cereals on the basis of an acknowledged experimental error . . . As far as Sharbati Sonora is concerned it was neither developed nor released, nor is it cultivated for its lysine content'.

Nevertheless, it would be fair comment to point out that Swaminathan's fault lay in the fact that although the reported value for lysine content was wrong, he allowed the erroneous figures to be used for more than a year. Perhaps the lesson to be learned is that enthusiastic senior scientists should be less eager in relating and publishing the most recent results of experiments, before they are properly confirmed. This 'custom' does not serve the interest of science and should, in my opinion, be censured.

Dr N. E. Borlaug, a Nobel Prize winner, and Dr R. C. Anderson, a famous wheat breeder, reacted to the accusations against Swaminathan in a letter to *New Scientist*:

> Dr Swaminathan, in our opinion, is one of the world's most effective agricultural scientists, educators and administrators. In the period from 1965–1972, while he was Director of the Indian Agricultural Research Institute (IARI), he was the organizing force and enthusiastic implementor in introducing the widespread cultivation of the high-yielding Mexican wheats. He also organized and introduced the improved technology which allowed these wheats to express their genetic potential in India. Wheat production rose from 10.5 million metric tons in 1965 to 26.5 million tons in 1972, a rapidity and magnitude of change in a major food crop that has been unmatched anywhere in the world. It was primarily on the basis of this contribution to this wheat revolution that Swaminathan was justly named recipient of the Magsaysay Award in 1971.[241]

Another remark in defence of Swaminathan, criticizing journalistic ethics was made by Kar:

> It is ironical that in 1974, when the world is threatened by one of the severest food shortages in history, and seven years after the unfortun-

ate reporting of one erroneous chemical analysis, that there appear to be those who would destroy one of the world's greatest and most productive agricultural scientists.[242]

Can Swaminathan's fault, in letting the erroneous results on the lysine content of his new breed of wheat be propagated for some time before admitting in public that they were wrong, be excused, or is this so serious a breach of scientific ethics that the verdict against him should be 'guilty'? I tend to support Borlaug's opinion that, in view of his significant contribution to agriculture, Swaminathan should be given the benefit of the doubt over the delay in correcting the erroneous information stemming from his research group. It is surprising, however, that a scientist in his exalted position should have so misjudged the ethical aspect of delaying for a year the revelation that the information propagated by his group was wrong. Sir Ralph Riley, retired Director of Plant Breeding Institute (Cambridge, UK) expresses the view (private communication) that the wrong result was due to 'inadequate supervision of the work of junior colleagues who may have found what the master wanted . . . The culpability relates to a man putting his name on too many papers deriving from work in which he had not participated'.

Compare here two situations. Which is unethical: to produce and publish results without doing the actual research (as was done, for example, by Burt) or to publish erroneous figures resulting from genuine and painstaking research (Swaminathan)? These are two different types of misconduct in science on very different ethical levels. The final verdict must be different in the two cases. In the first, even if guilt was admitted, the scientist would not be absolved of his unethical behaviour. In the second case, however, admission of error, *as soon as it was discovered* (note my stress) would have set the record straight without leaving a blot on the integrity of the scientist involved.

A FLUTTER IN THE DOVECOTE: THE ILLMENSEE CASE

Dr Karl Illmensee is Professor of Embryology and Developmental Biology of the Faculty of Science at the University of Geneva. At the end of a seminar of the Section of Biology at the university in January 1983, one of Illmensee's collaborators, Dr K. Buerki, announced that he and other colleagues could not accept Illmensee's results on successful transfer of nuclei from a mouse tumour into mouse eggs which had resulted in the formation of normal embryos. The implication was that Professor Illmensee had falsified the data. A flutter in the Geneva dovecote ensued.

On 17 May three professors at the Faculty of Science of the University of Geneva (Tissieres, Lämmli and Crippa) met Illmensee and confronted him with the accusations. At the end of the meeting Illmensee countersigned a declaration, already signed by the three others, stating: 'Protocols of experiments of Dr Karl Illmensee have been manipulated in a way which is contrary to scientific ethics in some period of 1982' (quoted in the report of the International Commission of Inquiry into the Scientific Activities of Professor Karl Illmensee, Geneva, 30 January 1984). The events seem worthy of our investigation.

In 1962 Briggs and King[243] reported that they had implanted a diploid nucleus from an embryonal cell of a frog into a fertilized but enucleated egg. The resulting amphibians were thus the descendants of only one parent, the one from which the diploid nucleus was taken. Illmensee, interested in this subject, subsequently spent some time at Jackson Laboratory in Bar Harbor, Maine, where he collaborated with Dr P. Hoppe in carrying out similar experiments on mice by replacing the nuclei of fertilized mouse eggs with diploid nuclei from developing embryos taken from a different strain of mice. The nuclei came from grey and agouti mice, while the recipient eggs came from black mice. The developing embryos should have been either grey or agouti. Sixteen of the operated eggs were implanted in the uteri of white foster mother mice together with 44 non-treated white mouse embryos to make up the litter size. Of 35 animals that were born of these embryos, 32 were white, two grey, and one agouti. The experiments thus demonstrated that transplanted eggs would develop normally. The publication of these results[244, 245] raised great interest in the scientific community, and many laboratories tried to repeat the Illmensee–Hoppe experiments, but without success.

In the meantime, Illmensee returned from Bar Harbor to Geneva, and there, in the spring and summer of 1982, carried out experiments on transfer of nuclei from mouse embryonal tumour cells (teratocarcinoma) into fertilized mouse eggs. It was against this series in particular that Dr Buerki and his colleagues levelled accusations of fraud. They challenged the availability of sufficient embryos for the experiments; they alleged that culture tubes were either empty or altogether absent from incubators during the period when the experiments of Illmensee were supposed to be in progress; they further claimed that Illmensee's protocol described an experiment purported to have been carried out on a certain Sunday when no appropriate tools appeared to have been used. Illmensee's report of these experiments, first at the Congress on Embryo Transfer in Mammals at Annecy, and later at the Geneva seminar, provoked the formal accusations that were the subject of the subsequently signed declaration of 17 May 1983.

At the Jackson Laboratory, too, a committee was formed to look for evidence of possible misconduct. This committee concluded that 'the failure to replicate did not necessarily mean that the results of the Illmensee-Hoppe experiments were invalid'[246]; (and also Editorial, *Science*, 1983, 220: 1023). The committee recommended, however, that Hoppe and Illmensee should repeat the experiments.

The Geneva allegations produced considerable turmoil both in Switzerland and abroad because of Illmensee's international reputation. The Rector of the University of Geneva appointed Professor Gardner of Oxford, Dr McLaren of London and Professor Chambon of Strasbourg to the international committee of inquiry. All were selected for their scientific expertise. Professors Heer, Martin-Achard and Renold represented the Faculties of Science, Law and Medicine of the University of Geneva. The commission conducted hearings and examined relevant documents and on 30 January 1984 issued a 28-page report. The report commented on the unacceptably large number of corrections, confusions and discrepancies in Illmensee's experimental protocols. Taking these into account, together with the seemingly secretive and non-communicative manner in which Illmensee had worked, it was perhaps not surprising, the report observed, that some of the Illmensee's junior colleagues were led to suspect that some or all of the experiments were contrived. However, while the accumulation of errors and discrepancies were such as to throw grave doubt on the validity of the conclusions, in the absence of more cogent evidence, the hypothesis that the protocols were fabricated could not be upheld.

The commission also examined an NIH application of Illmensee (of May 1983), entitled 'Gene expression during mammalian development'. On 18 May Illmensee sent a letter to NIH in which he withdrew a sentence relating to his alleged success in transferring teratocarcinoma nuclei into the cells of developing embryos resulting in the formation of mice with variegated coats made of sections partly derived from the genome of the teratocarcinoma cells and partly from the recipient mouse embryo cells. He stated in the letter that there had been a typographical error. Nevertheless, the committee deemed it possible that the original grant application 'contained an element of invention in respect to the existence of the male coat chimera' (p. 24 of the Report).

Both the Bar Harbor Committee and the International Commission recommended that the transplantation experiments be repeated before the possibility of cloning of mammals was accepted as an established fact.

On 29 May 1984, the NIH withdrew the $218,000 grant to Illmensee (*Nature* 1984, 309, 738).

In a letter to *New Scientist* of 7 June 1984 (p. 40), Illmensee rectified statements made about him in the periodical on 3 March 1984 (p. 7). He stated that he never refused to repeat the disputed experiments, that he had admitted to the inquiry committee that there had been errors in his experimental records, but that these errors did not affect the results of the experiments and that he had been reinstated by the university. Nevertheless, Illmensee resigned his post at the university, as of September 1984.

Among the ethical problems I discern in Illmensee's case is one relating to responsible reporting in the scientific press. Even before his scientific conduct had been officially investigated, and before the international commission had reached its conclusions, periodicals such as *Science, Nature* and *New Scientist* had brought the matter to the attention of the scientific and lay public. Thus, while the case was still *sub judice*, it was insinuated that Illmensee was guilty of misdemeanour. Once such an image has been built up, it is difficult for the accused scientist, if subsequently found innocent, to restore his reputation.

Reading the Committee's report one comes to the conclusion that Illmensee had not been cleared of some of the accusations. I stress, however, that even if he was guilty of misdemeanour, the verdict of the scientific press should have been withheld till *after* publication of the Report. Unethical behaviour by a scientist does not justify unethical treatment by the press.

The case also poses problems with regard to personal relations within a laboratory. In Illmensee's case 'the secretive way in which he worked and his failure to share techniques with his colleagues . . . provide an obvious basis for arousal of suspicion among those with whom he works' (Report of the International Committee of Inquiry into the scientific activities of Prof. Karl Illmensee, 30 January 1984,

p. 17). It should be the departmental chairman or faculty dean's business to know and to act when the relationship between a student and his mentor, or between a research worker and his supervisor, has become bad, as the investigation revealed was the case in this instance. If the differences between the two cannot be resolved, the student (or the worker) should be transferred from a situation that can be of little scientific and educational profit to him.

THE CONTROLLED DRINKING CONTROVERSY

A recent case of an accusation of fraud bordering on slander has been reported by A. G. A. Marlatt, concerning the controlled drinking controversy.[247]

The traditional view of alcoholism has been that it is a disease due to the loss by the drinkers of their ability to control their consumption of alcohol. According to Jellinck,[248] the loss of control occurs whenever the alcoholic consumes even a small amount of alcohol, which triggers the drinking behaviour. The cure therefore is total abstinence, and this has been the prevalent method of cure used by Alcoholics Anonymous.

During the past two decades studies have been published to show that alcoholism is a behavioural problem and not a disease. It can therefore be cured by individual behaviour therapy coupled with controlled drinking.[249-251] This therapy is based on measurement of blood alcohol levels coupled with aversive therapy, i.e., whenever the blood alcohol rises above a preset level, the patient receives a shock. A review of 22 studies by Miller and Hester[252] has indicated that in 21 cases the average success achieved was 65 per cent.

A detailed study by Mark and Linda Sobell,[253] working in the Addiction Research Foundation in Toronto (ART), indicated that a significant improvement of function had been achieved in a two-year follow-up of alcoholics who had received training in controlled drinking. This study was attacked by a group led by Mary Pendery,[254] who conducted a 10-year study of some patients originally treated by the Sobells. The group insinuated that there was doubt about the scientific integrity of the Sobells' research. One of the authors (Maltzman) was quoted in the *New York Times* of 28 June 1982 (p. A12) as saying: 'Beyond any reasonable doubt, it's a fraud.'

Such an accusation could not have gone unchallenged. The president of ART set up a 'Committee of enquiry into the allegations concerning Drs Linda and Mark Sobell', chaired by the Professor of Law and Criminology at the University of Toronto, Bernard Dickens, and composed of three other experts in medicine, psychology and administration, A. N. Doob, O. H. Warwick and W. C. Winegard.

On 5 November 1982 the committee issued a report. Having considered that the charges made and implied in respect to the Sobells involved most serious allegations, the report analysed more than 100 pages of Sobell's data and the accusations of the Pendery group, and reached the following conclusion:

> The Committee has reviewed all the allegations made against the Sobells by Pendery et al. in their draft manuscript, in their published *Science* article and in various statements quoted in the public media. In response to these allegations, the Committee examined both the published papers authored by the Sobells as well as great quantity of data which formed the basis of these published reports. After isolating each of the separate allegations, the Committee examined all of the available evidence. The Committee's conclusion is clear and unequivocal: The Committee finds there to be no reasonable cause to doubt the scientific or personal integrity either of Dr Mark Sobell or Linda Sobell.

Since the Sobell's research was supported by the NIH, their work was also investigated by the Committee on Science and Technology of the US House of Representatives. The chairman of the Subcommittee on Investigations and Oversight, James E. Jensen, exonerated the Sobells completely.

It was thus clear that the Sobells were unjustly accused. Nevertheless, they found it necessary to publish a detailed point-by-point refutation[255,256] of Pendery's group's accusations. The Sobells showed that the experimental and control groups were comparable in terms of pretreatment characteristics, that they were real alcoholics, that the assignment to groups was random and that the mortality in the experimentally treated group after 10–11 years was 20 per cent, while in the traditionally treated group it was 30 per cent. In short, the Sobells demonstrated that their accusers' article rather than being a refutation of their findings, actually strengthened the validity of their report and conclusions. The attack on the Sobells reflects a scientific revolution in progress.

Returning to the subject of controlled drinking, D. L. Davies, the British physician and alcoholism researcher who initiated the controlled drinking method, wrote:

Yet so strong are entrenched ideological views on this issue, that the argument waxes even more fiercely, recalling the 19 century battles between wets and drys, using indeed the very language and thoughts of early 19 century temperance workers with the same preoccupation with the morals and religious aspects of the 'first drink' and the role of divine help.[257]

Although today there is still lack of good empirical support for the effectiveness of any type of treatment for alcoholism, the proponents of the disease model insist that alcoholism is a progressive disease that can be only temporarily arrested by total abstinence. (Drinking is a sin, and only abstention offers salvation.) Scientists who suggest that controlled drinking may offer a cure 'are branded as agents of the devil',[247] and therefore their experimental data must be a lie.

The debate rages on an emotional basis. Here it seems the determined beliefs in the correctness of a hypothesis (that is, total abstinence is the only cure for alcoholism) led a research group to accuse the proponents of an alternative idea of fabricating their results. The views of traditional, normal science did not conform to new experimental evidence and thus it was that the Sobells came under violent, but unsuccessful attack.

CELLS THAT WERE NOT WHAT THEY SEEMED

In a paper published in 1980, Dr Nelson-Rees of the University of California Naval Bioscience Laboratory reported that a line of cells, named T-1, which had been used in many laboratories as a normal kidney cell line, was in fact a descendant of HeLa cells.[258]

The T-1 cell line was originally set up in Holland in 1957, and from there it was sent to other laboratories, each continuing to propagate the cells in culture. Four of these laboratories sent their T-1 cells to Nelson-Rees and he found them to be contaminated with HeLa cells.

Determination of the identity of a cell line has been made possible by the study of their karyotypes, that is, the number, shape and distribution of the chromosomes, as well as by the study of enzymes present inside the cells which may serve as 'fingerprints'.

Nelson-Rees's findings turned the attention of radiologists and cytologists to a paper published ten months earlier by Paul S. Furcinitti and Paul Todd of Pennsylvania State University.[259] They described their measurements of the minimal dose of ionizing radiation on T-1 cells that would cause genetic damage to these cells, assumed to be normal kidney cells when, in fact, they were cancer cells. The important question is, does it really matter on which human cells one measures the radiation dosage leading to the death of the irradiated cells (as determined by the slope of the line relating radiation dose to cell viability after irradiation), or is it important whether there is a threshold below which the cells would not be damaged?

Opinions on this matter vary among scientists. Nevertheless, the argument against Todd and Furcinitti was that they had known since 1977 that the T-1 line they worked with was probably contaminated with HeLa cells, because they had been notified of this fact by the American Type Culture Collection where Todd had wanted to deposit the T-1 line. The implication was that they had published false data.

Ironically, as it turned out,[260] Todd tried in all his papers submitted for publication since 1977 to warn that the real identity of T-1 cells was doubtful. Unfortunately, reviewers, like the reviewer of Todd's 1978 paper,[261] demanded that the passages raising doubts about T-1 and suggesting their HeLa origin be deleted. One reviewer wrote that in order to reach therapists 'the manuscript needs to be simplified and details omitted'.

The ethical problem in this case is whether the editors should have the right to demand from the author that data which he considers pertinent be removed from a paper, or whether an author should absolutely insist that facts which he considers 'potentially troublesome' should be published, or else withdraw the paper if the editors insist on omission of these details. In Todd and his co-workers' case it is clear that he kept pointing out the problem of cell identity in his manuscript, but he did not consider it important whether the cells he studied were malignant or not, assuming that the sensitivity to gamma radiation of these cells would be the same.

I certainly do not agree with Broad's statement that 'significant is the fact that the Penn. State researchers [Todd and co-workers] for more than three years after being alerted to the possibility that theirs was a malignant line, took no action to pin this possibility down', which is laid as an accusation of wrong-doing by Todd and Furcinitti. At most they could have been guilty of an error of judgement. In my opinion their integrity should certainly not have been called into question publicly.

QUESTIONABLE NUTRITIONAL RESEARCH

The last example involves 'irregularities' discovered in the nutritional research of William E. Wheeler.

Those involved in livestock research and production have always sought to improve the digestibility of rations in ruminant animals. Of the minerals in animal feed, the most important is calcium, the source of which is limestone. The effects of various limestone sources on the digestibility of rations have been studied in order to detect the most beneficial effects of beef rations.

In view of these concepts, nutritionists accepted as logical and credible the findings of Wheeler, published from the Ralston Purina Company in St Louis, Missouri. With his colleagues, Wheeler[262-263] reported on the effects of various types of limestone, differing in particle size and reactivity rate. Research of other nutritionists, however, failed to confirm Wheeler's findings.

Dr Bob Oltjen, the director of the Meat Animal Research Center at Clay Center, Nebraska, discovered some troublesome discrepancies in Wheeler's research, and in January 1982, together with Dr William Pond, issued the following statement:

> Dr William E. Wheeler has recently resigned from his position at the Roman L. Hruska US Meat Animal Research Center, Neb. Readers should be aware that several irregularities had been discovered in some of his research data dealing with cement kiln dust and limestone levels and sources in ruminant diets. It is emphasized that these irregularities do not invalidate the concept of buffering capacities nor the beneficial responses reported in the scientific literature by others. However, the consistency and magnitude of the responses reported by Dr Wheeler should be scrutinized and carefully interpreted.[264]

Wheeler himself published an erratum statement in the *Journal of Animal Science* and *Journal of Dairy Science*:

> Irregularities have been discovered in data reported in abstracts and manuscripts published in the Journal of Diary Science on which W. E. Wheeler is senior author. Therefore the validity of the reported information is questionable. (Quoted by Eng.[264])

In spite of these statements, Kenneth Eng, still believes, and is supported in this belief by other nutritionists, that higher calcium and limestone levels have some merits and lead to better utilization of starches in food.

I am unable to say what Wheeler's 'irregularities' signify and therefore am left only with the fact that he admitted them, and thus undermined the validity of his research results.

POSTSCRIPT

As a postscript to this chapter, I relate here a short anecdote concerning Roentgen and Lord Kelvin.

When Roentgen published his discovery of X-rays, Lord Kelvin pronouced this an elaborate hoax. He thought that the production of new rays by a cathode ray tube was unbelievable; after all, cathode ray tubes had been in use for ten years before Roentgen's discovery.

Presumably what this meant to Kelvin was that many famous physicists of that era who were experimenting with cathode ray tubes had been producing X-rays without knowing it. In this situation it seems, to excuse collective blindness, one accuses the discoverer of a new phenomenon of perpetrating a hoax.[265]

9

Forgeries in Palaeontology
and Archaeology

THE PILTDOWN HOAX

One of the most elaborate scientific hoaxes ever perpetrated was the so-called 'Piltdown Man': the fraud remained undetected for some 40 years. The story, was vividly and critically depicted by J. S. Weiner in his book *The Piltdown Forgery*,[266] written soon after the scientific exposure of the hoax in the *Bulletin of the British Museum* on 21 November 1953.

Officially, the story started in December 1912, when Arthur Smith Woodward, the Keeper of the Department of Geology at the British Museum, and Charles Dawson, an amateur geologist, announced at a meeting of the Geological Society in London that they had found the missing link between the apes and man. Their findings in ancient gravels of the Sussex Ouse (at Barkham Manor) consisted of a few pieces of cranium of an essentially human skull, a piece of mandible which appeared to be that of an ape and which had some well-worn molar teeth, some fragments of teeth of hippopotamus, deer and beaver, as well as some ancient remains of elephant, mastodon and rhinoceros. This unique find of a human brain case and an apish jaw represented an evolutionary link that fitted Darwin's theories, and therefore could be expected to exist by the educated public. Quite understandably it created something of a stir in the anthropological world. 'That we should discover such a race at Piltdown, sooner or later, has become an article of faith in the anthropological creed ever since Darwin's time'.[267]

Woodward and Dawson's interpretation that the components of Piltdown Man (named *Eoanthropus dawsoni*) were parts of the same individual, dating back to the early Ice Age, was not unanimously accepted. David Waterston, at Burlington House (Royal Academy), thought that the cranium and the jaw belonged to different creatures, and in spite of additional evidence to the contrary, he (and others) held this view until his death in 1921.

During the next few years various finds connected with the Pilt-down Man were reported. In 1913 the Roman Catholic priest and amateur palaeontologist Pierre Teilhard de Chardin, who took part in the first discovery, found in the same gravel bed a worn eye-tooth (canine tooth) that fitted the mandible found previously. In 1914 workmen digging in the same bed uncovered an elephant bone which had been shaped to form a club-like tool. The matter of the identity of the Piltdown Man was clinched in 1915, when Dawson himself reported finding remnants of another Piltdown individual at Sheffield Park, some two miles away from Barkham Manor. He also found there a rhinoceros tooth stemming from the lower Pleistocene era.

In 1916 Dawson fell ill and died. Following his death all the ex-cavations at the site of the Piltdown findings were unsuccessful, and nothing more was found there.

At the time of the Woodward–Dawson discovery the only other remains of the evolving human race were the Java Man, found in 1891 by Dubois, and the Heidelberg jaw found in 1907 by Maurer. It was only in 1936 that new finds in Java confirmed Dubois' claim that *Pithecanthropus erectus* was a primitive hominid. Similar creatures were also found in that period in Peking – *Sinanthropus pekinensis* – and later Dart and Broom found the ape-looking, but incipiently human *Australopithecus* in the Transvaal. These later finds contra-dicted the evolutionary trend as expressed by the *Eoanthropus dawsoni* which had a human cranium and an ape jaw; the Java and the Peking finds, in contrast, had a simian-looking forehead, but their jaw was very different from that of apes. In order to reconcile the Piltdown with the Java and Peking finds, Woodward postulated in 1944 that these might have been two different evolutionary lines, and the matter rested there until about 1952.

At the time of the Piltdown find, and for the next 37 years, the bones found were considered to belong to the early Ice Age (Pleistocene), in accordance with the antiquity of the animal bones accompanying the Piltdown skull. More precise dating methods were developed by Dr Kenneth Oakley, who adapted the method of a French mineralogist, Carnot (1892), to determine the age of the bones by their fluorine content. (The amount of fluorine in fossil bones increases with their geological age.) Oakley found that the content of fluorine in the mastodon and elephant bones from Piltdown was about 2 per cent, while that in the fragments of the Piltdown jaw was only 0.1–0.4 per cent.[268] This determination put the age of the Piltdown Man to the upper Pleistocene era, that is, it was about 50,000 (rather than the previously estimated 500,000) years old. The Piltdown Man was thus not a link, but rather an isolated evolutionary branch, and therefore of much lesser importance than previously believed. In fact, at a congress of palaeontologists in July 1953, the discussion revolved around Java, Rhodesia, Neanderthal and Transvaal Man; Piltdown was not even mentioned.

During dinner at that conference, Drs Oakley, Washburn and Weiner pondered what had happened to the Piltdown theory. They raised the suspicion that actually there had been too many things wrong with the Piltdown Man. They noticed that there was no record of the site where the second group of finds (Piltdown II) were obtained and there was something wrong with the teeth – the molars and the canine were worn flat, as indeed occurs in aged humans, but the wear was not related to the physiological age of the teeth, and they looked as if they had been mechanically filed down; the dentin underneath the stained surface was pure white.

These suspicions led the scientists to approach the British Museum with a request to renew the investigation of the Piltdown material by anatomical, radiological and chemical methods. The results of these tests, published first by Weiner, Oakley and LeGros in 1953,[269] clearly indicated that the Piltdown discovery was a hoax.

What was the evidence? The test for the content of fluorine and organic material showed that the jaw was modern and of a different origin to the cranium; the staining of the pieces of cranium was brown throughout, while the stain on the jaw was superficial. The stain in the canine teeth was identified as Vandyke brown; X-ray analysis of the cranium indicated the presence of sulphate in it, while there was none in the jaw, which was stained by chromate. The abrasion of the teeth was unusual in that the borders of the worn-down surfaces were sharp and not, as would be expected by natural wear, bevelled; the X-ray examination confirmed that the jaw was of a modern age, with the teeth treated so as to resemble human ones. Actually the jaw was shown to be identical to that of an orangutan. All this evidence showed that the cranium and the jaw could not have belonged to each other, and that their presence in the same geological layer was quite incompatible. The Piltdown Man was thus an artefact in that it was composed of a probably genuine cranium and a fraudulent jaw.

If the Piltdown Man was a forgery, who was the 'forger'? In addition to Woodward and Dawson, there were at least two other people involved who might have perpetrated the hoax or assisted in it.

Woodward, the Keeper of the Department of Zoology at the British Museum (as well as the Secretary and President of the Geological Society) specialized in research on fossil fish and reptiles. He was considered by his contemporaries to be a man of highest integrity.

Dawson, who knew Woodward for some 20 years and was on friendly terms with him, was an energetic, genial and enthusiastic person. He was by profession a lawyer and antiquarian, deeply interested in geology and archaeology. At the early age of 21 his works brought him the Fellowship of the Geological Society and later the status of Honorary Collector for the British Museum. Dawson found a new species of Iguanodon, which he named after himself, and made numerous other geological discoveries. He wrote a classic book on Hastings Castle (see below pp. 149–50). His law office at Uckfield was quite successful; in 1905 he took a partner and after that devoted most of his time to his hobbies.*

After Dawson's death, Woodward said:

> He had a restless mind, ever alert to note anything unusual, and he was never satisifed until he had exhausted all means to solve and understand any problem which presented itself. He was a delightful colleague in scientific research, always cheerful, hopeful and overflowing with enthusiasm.[270]

In February 1912 Dawson wrote to Woodward that he had 'come across a very old Pleistocene bed overlaying Hastings beds between Uckfield and Crowborough which I think *is going* to be interesting', and then reported that he found there parts of Pleistocene human skull. In June, Dawson, Woodward and Teilhard de Chardin went

---

* It might be of significance that Dawson found an unusual human skull at Barcombe Hills in 1906. This fact was mentioned in the *Sussex Express* of 1 January 1954): 'Mrs Florence Padham . . . remembers that in 1906, when she was aged 13, and living in Victoria Cottage, Nutley, her father gave Charles Dawson a skull brown with age, no lower jaw bones, and only one tooth of the upper jaw, with a mark resembling a bruise on the forehead. Dawson is supposed to have said: "You'll hear more about this, Mr Burley".'

to the spot where the cranium pieces had previously been located, and there they found the jaw.

Of the trio involved in finding the Piltdown fossil, a special role is assigned to Pierre Teilhard de Chardin, who worked on the search and excavations with Dawson from 1908 until 1914. Teilhard was born in France and studied for three years on the Channel Island of Jersey. He then taught physics and chemistry in Cairo. From there he came to Hastings in 1908 to finish his theological seminary and was ordained in 1912. During the hunts after fossils with Dawson, it was Teilhard who found the first Stegodon tooth and a carved flint tool, and the next year it was he who found the canine tooth that fitted the mandible. With the outbreak of the First World War Teilhard joined the French army in France as a stretcher bearer, and stayed there until the end of the war. In 1922 he became Professor of Geology at the Catholic University in Paris, and as such went to China. There, in 1923, he established the existence of a palaeolithic man in Northern China. He remained in China for most of his life doing research in palaeontology and writing philosophical books on cosmic theory and the relations between science and religion: his most famous work is *The Phenomenon of Man*. Teilhard de Chardin died in 1955.

Though Weiner regards Teilhard as an 'unwitting dupe' in the Piltdown affair, evidence accumulated after Weiner's book had been written, led Stephen Jay Gould, Professor of History of Science at Harvard, to a hypothesis that Teilhard was the person mainly responsible for the forgery, in that he was the one who planted the fake evidence.[271, 272] Gould believes it was probably meant to be a prank, rather than a malicious forgery, but it went too far. Gould supposes that Teilhard de Chardin, who at the time was as a 30-year-old theology student with a sense of fun, spent a lot of time with Dawson and together they may have concocted the plot. But why? Dawson's motive perhaps was to expose the gullibility of the academic scientists, and Teilhard, the Frenchman, maybe wanted to make fun of the English who still had no legitimate human fossils.

Teilhard left for France at the end of 1914, therefore Dawson's second find at Sheffield Park in 1915 could not have involved Teilhard directly. Why then should he be implicated? Gould reviewed the correspondence between Teilhard and Oakley (the scientist who carried out the fluorine studies and showed that the cranium and the mandible did not belong together). In a letter Teilhard stated that Dawson was with him at locality II (Sheffield Park), and showed him the pieces of skull and an isolated molar from there. This could have occurred only in 1913, when Dawson and Teilhard were known to have been at Sheffield Park and when no findings were reported. Dawson reported the finds to Woodward in January 1915 (cranial pieces) and July 1915 (the molar tooth), but at that time Teilhard was in France and could not have seen these pieces. Teilhard's admission to Oakley of having seen these bones during the visit to the site, is therefore an admission of complicity.

Gould stresses that though Teilhard played an important role in establishing the existence of *Eoanthropus*, there are very few statements or remarks on the Piltdown find in all his scientific writings relating to the antiquity of man. This, and other circumstantial evidence collected by Gould from Teilhard's colleagues, led Gould to conclude that the forgery was probably a joke perpetrated by Teilhard and Dawson. Because of the outbreak of war Teilhard did not have the opportunity to put the record straight, Dawson died, and by that time the Piltdown story was well established and supported by many important men of science; it would have been extremely damaging to Teilhard's career to have to admit now that he was involved in the hoax.[270, 271]

Another possibility examined by Gould concerned tape recorded evidence given on his deathbed by J. A. Sollas, an Oxford palaeontologist who stated that he disliked Woodward so much that he set him up to perpetrate the hoax to demonstrate Woodward's ignorance in human palaeontology. One of Sollas's previous tricks on Woodward was letting him describe an engraving of a horse head on an ancient bone as genuine, when Sollas knew that it had actually been made by students who copied another ancient drawing.

Another figure, who according to Weiner might have played a prominent role in the affair, was Lewis Abbot, an amateur geologist and jeweller from Hastings. He knew Dawson well as in his discussion with him and with others Abbot claimed that Pliocene formations should be recognizable in that part of England. Abbot found numerous vertebrate species in the Shode Valley in Kent, and was a recipient of the Lyell Award of the Geological Society. He had a large collection of flints, animal bones and ape remains, and was always discussing his discoveries. Abbot wrote to Woodward in December 1912 that Dawson would not have made the discovery of the Piltdown Man but for his (Abbot's) inspiration and instruction.

While Weiner, therefore, is inclined to believe that the person most likely to have been responsible for the forgery was Charles

Dawson ('. . . it is not possible to maintain that Dawson could not have been the actual perpetrator; he had the ability, the experience, and . . . he was at all material times in a position to pursue the deception throughout its various phases'), Gould maintains the opinion that Teilhard de Chardin was a partner to the conspiracy.

Additional evidence implicating Dawson comes from information gained by Weiner from A. P. Pollard, Assistant Surveyor of Sussex County Council. He stated that he had been friendly in 1926 with Harry Morris, a bank clerk and amateur archaeologist who lived in Lewes at the time of the Piltdown affair. Morris had a collection of flints which Weiner was able to inspect. One flint in the collection bore an inscription: 'Stained by C. Dawson with intent to defraud (all) – H. M.', and a note saying that the stain was potassium permanganate, and that he got the flint from Dawson. In another note in the flint collection Morris wrote. 'I challenge the SK [South Kensington] Museum authorities to test the implements of the same patina as this stone which the impostor Dawson says "were excavated from the Pit!" ' They will be found to be *white* if hydrochlorate acid be applied. H. M. Truth will out.' A still different note carried the message that the 'canine tooth was imported from France' (Teilhard?).

Also living in the area at the time was the Governor of Lewes jail, a retired major, R. A. Marriott. He was a friend of Morris and was also convinced that 'Piltdown Man' was a fraud. After the disclosure of the forgery in 1953, Mrs Olivia Lake, Marriott's daughter, remembered her father saying that the jaw and the canine tooth of Piltdown were faked. Another contemporary who was convinced of fraud was Captain Guy St Barbe, Keeper of Geology at the British Museum. In 1912 he lived near Lewes, where he met socially with Dawson. On one occasion, when visiting Dawson's office, he found Dawson staining some bones in a vat. He had explained with embarrassment that he wanted to find out how the bones became stained in nature. A friend of both Morris and Abbot, A. S. Kennard, a known palaeontologist of the Natural History Museum in London, also believed that Piltdown was a hoax and claimed he knew who the perpetrator was, but he died in 1948 without disclosing a name.

In 1981 suspicion fell on yet another person as being involved in the hoax. He was Martin A. Hinton, a zoologist from the British Museum.[273] He visited the site of Dawson's primary discovery in 1913, and had an ape tooth, filed down to make its surface smooth, buried at the site. It is supposed to have been he who took a leg bone of an elephant from the museum's collections, carved it to make it look like a cricket bat and buried that too. Matthews believes Hinton did this to get even a bigger laugh when the hoax was exposed, but unfortunately for him, all those involved took even Hinton's artefacts seriously as a part of the whole Piltdown picture.

Summing up his research on the affair, Weiner stated:

> As long as the weight of circumstantial evidence is insufficient to prove beyond all reasonable doubt that it was Dawson himself who set the deception going by 'planting' the pieces of brain case, our verdict as to the authorship must rest on suspicion and not proof.

It is possible, muses Weiner, (and in this he is supported by Gould), that Dawson was implicated in a 'joke', perhaps not even his own, which went too far.

The amazing fact is that from the start the true explanation of the nature of Piltdown Man was available, but hope, cultural bias and prejudice prevented scientists from seeing and accepting it. In the early 1940s a famous human anatomist, Franz Weidenreich, wrote about the Piltdown Man: '*Eoanthropus* should be erased from the list of human fossils. It is an artificial combination of fragments of a modern human braincase with orang-utan like mandible and teeth.'[272] To this came the ironical reply of Sir Arthur Keith: 'This is one way of getting rid of facts which do not fit into a preconceived theory: the usual way pursued by men of science is, not to get rid of facts, but to frame a theory to fit them.'

Why did the English palaeontologists accept the Piltdown man so easily? Scientists, contrary to lay belief, do not work by collecting only 'hard' facts and fitting together information based on them. Scientific investigation is also motivated by pursuit of recognition and fame, by hope and by prejudice. Dubious evidence is strengthened by strong hope: anomalies are fitted into a coherent picture with the help of cultural bias (in the case of Piltdown, the preconceived notion was that the evolution of man had been conditioned first by the development of a large brain, a belief which was later found to be completely unfounded). Such a notion made easy the acceptance of a large cranium with an ape-like jaw. On this basis it may be easier to make a discovery that is predictable by the current scientific notions, than one which is completely unexpected.

Why did the forgery remain undetected for 40 years? One reason, in addition to those mentioned above, was that the British Museum, where the Piltdown pieces were stored, made access to them extremely difficult: scientists were permitted only visual inspection of the actual remains and handling was restricted to replica casts of the skull and mandible.

It seems proper to end this story with a quotation from Darwin's *Descent of Man*: 'False facts are highly injurious to the progress of science, for they often endure long; but false views, if supported by some evidence, do little harm, for every one takes delight proving their falseness.'

DID ARCHAEOPTERYX EXIST?

A suspicion of another great palaeontological fraud has recently been raised; it concerns the authenticity of the famous fossil Archaeopteryx, thought by some to be the ancestor of birds.

The discovery of *Archaeopteryx litographica* had been hailed in the nineteenth century as evidence for the missing link between reptiles and birds, as a proof of Darwin's theory of evolution. Archaeopteryx differed from the then known flying reptile, Pterdodactyl, by being equipped with real feathers.

The first specimen of fossilized Archaeopteryx was found in a Bavarian quarry near Pappenheim by a German physician, Dr Karl Haeberlein, in a layer of limestone 160 million years old (Upper Jurassic). Haeberlein sold the rock with the fossil to the British Museum for £700. This specimen is composed of two halves created when the rock containing the fossil bones and feathers was split, like two halves of a mould. One can distinguish there the skeleton of an animal with wing and tail feathers, but no head.

The possible existence of such a feathered reptilian bird was predicted before 1861 by Thomas H. Huxley in a sketch showing the animal with wings, scales on the body, a lizard-like tail and teeth in the 'beak'.

In 1877 Haeberlein's son, Ernst, found another specimen in the vicinity of the location where the first specimen originated. This Archaeopteryx was well preserved and had a head and teeth. Haeberlein sold it to Werner von Siemens (the founder of the great electrotechnology industry) for 20,000 gold Marks (about £1000) and from him the fossil reached the Humboldt Museum in Berlin, sold for £1000. Later, in 1951, 1956 and 1970, three additional specimens were discovered in various locations and found their way to museums in Maxberg, Eichstadt and Haarlem.

For almost 100 years the Bavarian fossils proudly attested to the evolutionary transition from reptiles to birds, until in 1985 a group of scientists composed of Sir Fred Hoyle and Chandra Wickramas-inghe, astrophysicists, and Robert Watkins and Lee M. Spetner, physicists, raised the suspicion that the Archaeopteryx fossil was a fraud.[274-276] They inspected and photographed the specimen in the British Museum and came to the conclusion that the impression of feathers in the rock was not genuine, but that it had been added to the original reptilian skeleton.[277, 278]

According to Spetner, chicken feathers were pressed on to a cement mixture made with powdered limestone originating in the same quarry so as to make an impression as if the feathers had been growing out of the wings and tail. The photographic techniques employed by Walker (low angle tangential flash lighting from different directions) revealed fine structural details on the fossil and the rock indicating a possible forgery. Indeed, inspection of the photographs indicates that the material in which the feathers are impressed is much finer grained than the underlying rock. In addition, one can distinguish some elevated finely grained spots or blobs on one slab while there is no corresponding cavity on the counter slab.

At the time the first Archaeopteryx was discovered, a very fine and detailed drawing of it was made by R. Owen in 1862.[279] When the recent photographs were compared with this drawing, unexplained discrepancies in the area at the end of the feathers were observed.

The evidence of Hoyle and his colleagues that the Archaeopteryx fossil had been manipulated (presumably by the Haeberleins) has been contested by some palaeontologists, among them Professor Ostrom from Yale University. He points out that out of the five existing specimens, three were found in this century by different people in various locations, all traceable to Upper Jurassic. This argument is countered by the accusing scientists, who state that only the two Bavarian specimens had real feathers, while the other fossils do not possess any unequivocal impressions of feathers and seem therefore to be only reptiles. They also point to the fact that the direction of the hip bones in the Haeberlein specimens corresponds to that found in birds, while in the other specimens it corresponds to reptiles.

So, is Archaeopteryx a genuine fossil or a fraud? Palaeontological and archeological forgeries were quite 'popular' in the eighteenth and nineteenth centuries and there is good evidence that excellent specialists existed in this field; Leonard Barth, for instance, the owner of Oehningen limestone quarries on the shores of Lake Constance, about whom the custodian of the Teyler Museum in Haarlem, T. C. Winkler, said: 'The keenness, the inventiveness and the artful ingenuity of the men of Oehningen knows no bound!'.[280]

In the case of the Piltdown skull, discussed previously, almost 40 years elapsed before the fraud was exposed by careful chemical analysis of the jaw and skull. No such fine chemical analyses of Archaeopteryx have yet been reported. One has therefore to suspend judgement, until all the required physiochemical tests have been performed, which will prove whether the feather imprints are 160 million years old, or stem from the avaricious imagination of a nineteenth century doctor.

MORE NOTORIOUS FORGERIES

In the nineteenth century the field of archaeology was plagued by a number of frauds. Their exposure was not easy, because of the small number of experts working in the field. While in the Piltdown case amateurs and scientists were equally involved, a number of earlier archaeological forgeries were perpetrated by skilful artisans and antiquity hunters and dealers.

A notorious case of forgery involved an antique dealer from Jerusalem, one Moses Shapira. Originally a Jew, he converted to Christianity and took the name William Benedict.

In 1897 Shapira offered for sale to the Berlin Museum 15 strips of parchment scrolls containing the original text of the Book of Deutcronomy. He informed the curators of the museum that he had obtained the scrolls from a bedouin who had found them in a cave in the gorge of the river Arnon, on the eastern shore of the Dead Sea. The Berlin Museum was not interested in the scrolls for good reasons. Some ten years earlier it had purchased from Shapira several hundreds of clay figurines, some claimed to be of ancient Moabite origin, and all later found to be modern artefacts.[281] After the Berlin refusal, Shapira whent to London and offered the scrolls to the British Museum for £1 million. The scrolls were examined there by C. D. Ginzburg, museum's adviser on ancient semitic scripts. Ginzburg published an article in *The Times* describing the scrolls and their importance. Two strips of the scrolls were exhibited to the public in the museum.

At that time the French archaeologist Charles Clairemont-Ganneau, who was employed in the French consulate in Jerusalem, arrived in London. He had attained fame as the discoverer of the Moabite Mesha stone describing the battles between the Moabite king and the Judaean and Israeli kings and corroborating the biblical story (Kings, 2: 3). Clairemont-Ganneau was also known for having found in Jerusalem a Herodian tablet with a Greek inscription prohibiting the entrance of Gentiles on to the temple mount.

During his visit to London, Clairemont-Ganneau examined the scrolls exhibited at the British Museum and declared them to be strips cut off a large Bible scroll which Shapira had sold to the British Museum a few years earlier. Confronted with this information, Ginzburg re-examined the scrolls and confirmed the Frenchman's opinion. Shapira took back the scrolls, went to Rotterdam and there committed suicide. The scrolls that were in his possession disappeared.[282] In view of the later Qumran finds of Dead Sea scrolls, which were authenticated, some modern scholars now entertain a view that Shapira's scrolls, allegedly coming from the same area might have been genuine after all.

Among other famous archaeological frauds one may mention the crown of Saitafernes, a fifth century Scytho-Greek work of art. As it turned out, it was made in the last quarter of the nineteenth century by an artisan from Odessa, Israel Ruchomovsky. The crown was sold to the Louvre in 1895 as a genuine fifth century object. Ruchomovsky later admitted to have forged the object.

Another celebrated forgery was that of an Etruscan sarcophagus 'found' in 1927 in the town of Glozel in the south of France. On the lid of the sarcophagus there were the reclining figures of a husband and wife. This piece was considered to be a part of the 1873 find from an Etruscan burial ground at Cerveteri in Italy, and a classic example of Etruscan art. Only a decade after the Glozel find, it became known that the sarcophagus had actually been produced by the brothers Penelli in response to the great demand for archaeological objects at the end of the nineteenth century.

One of the earlier documented cases of forgery comes from the eighteenth century. In 1724 Johannes Bartholomeus Adam Beringer, Professor of Natural History at Wuerzburg University, discovered in the shell limestone formation near Wuerzburg a number of stone figures representing a variety of animals and plants. He was guided to these finds by three peasant youths led by Christian Zaenger. Altogether some 2000 of these stone figures were excavated. Some of their finds the boys sold to the Prince–Bishop of Wuerzburg and to other members of the local gentry. In 1726 Beringer published a book, *Litographiae Wircburgensis*, with pictures of 200 of these stones.

Soon after publication of the book a scandal exploded. The three youngsters were brought to court charged with having falsified, buried and sold the stone figures. Zaenger, who was the chief culprit, confessed that he had been put up to the forgery by two professors of the university, Johann Georg von Eckhart, a historian and librarian, and Ignaz Roderique, Professor of Geography and Algebra. Eckhardt was known as a scholar of prehistoric studies in Germany, but also as an opportunist (he converted from the Protestant to the Catholic religion in order to get a position he wanted at a Catholic university).

Because of rather strained relations, Eckhart devised the scheme to ridicule Beringer, who was an ardent collector of fossils. He even went so far as to aid Beringer during the excavations and vouched in public that Beringer had indeed excavated the stone figures. It was Roderique who taught Zaenger the technique of producing forged figurines. The youths buried them in the limestone formation and led Beringer to the 'treasure'. When the identity of the instigator of the forgery was revealed in court, Eckhardt was dismissed by the Prince and retired to private life. Roderique left Wuerzburg and became a journalist. The Prince, who acquired a number of the stone figures, had all the documents pertaining to the case sealed; they came to light only after 200 years. As to Beringer, in spite of having been made a laughing stock in the academic circles in Wuerzburg, he still believed until the end of his life, 14 years later, that most of the figures he had were genuine.

A most recent exposure of an artistic forgery concerns the 'Rospigliosi cup', a Renaissance work attributed to Benvenuto Cellini the Florentine sculptor and goldsmith of the sixteenth century.[283] The curator of London's Victoria and Albert Museum, Charles Truman, found in the museum in 1929 a cache of some 100 drawings made by a nineteenth century German goldsmith, Reinhold Vasters from Aachen. Some drawings depicted Renaissance objects, others were working drawings for making new pieces. Upon publication of Truman's findings in *Connoisseur* implicating Vasters as Cellini's forger, the Metropolitan Museum of Art in New York examined its Renaissance collection and found that some 30 pieces of that collection, including the Rospigliosi cup, were Vaster's forgeries.

10

Plagiarism or Piracy?

It is very difficult to estimate the extent of plagiarism in science, though the insinuation of plagiarism and the theft of ideas is often encountered among scientists. The theft of unpublished findings or of data is, of course, much easier than that of published material, but examples of both types of misdemeaniour are well known in scientific circles.

According to Hagstrom,[284] out of a series of 1309 academic scientists in the exact sciences, 25 per cent complained of having their ideas 'stolen' by others, or at best, of their ideas not having been acknowledged.

For a crime to be perpetrated there has to be an incentive and an opportunity. In the case of plagiarism the incentive is clear and need not be expanded. Opportunity is provided by the intrinsic structure of scientific controls: it is provided by the institution of the peer review system, under which scientists receive for evaluation and criticism the applications of other scientists for grants, or manuscripts submitted for publication. It demands great integrity on the part of the reviewer not to make use of information gleaned from a grant application or from an as yet unpublished article, if such information is useful in one's own research. The danger here is real, since the referees and members of 'peer review groups' are chosen from a list of experts working in the same field. Moreover, plagiarism need not be conscious. People sometimes wrongly attribute ideas to particular scientists or forget their source and claim them as their own. Referees and reviewers must always be on their guard against this trap. It is now not unusual for authors who feel their work may have commercial applications to ask editors *not* to send their manuscript for review to named potential competitors.

Maddox, in an article published in *Nature* (1984, 312: 487) stressed the need for maintaining privacy within the peer review system.

One of the most notorious cases of extensive plagiarism recently discovered was that of Elias A. Alsabti.

In April 1980 the *Lancet* published a letter by Dr E. Frederick Wheelock[285] which related to two reviews published by Alsabti in 1979 in *Journal of Cancer Research in Clinical Oncology*[286] and in *Neoplasma*.[287] Wheelock found these two reviews to be identical, and moreover, two-thirds of the content consisted of an almost verbatim copy of a research grant application entitled 'Tumor dormancy and emergence' which Wheelock had previously submitted to the US Public Health Service. Wheelock also discovered that the rest of Alsabti's article came from some early drafts of Wheelock's manuscripts. The explanation for this amazing similarity between Wheelock's manuscripts and Alsabti's review papers could only be explained by the fact that Alsabti had spent five months in Wheelock's laboratory and had had access to the documents in question without Wheelock's knowledge or permission.

Soon after the publication of this letter a number of other articles appeared[288, 289] exposing further examples of plagiarism by Alsabti. One widely publicized case of plagiarism by Alsabti was his paper on mutagenesis by platinum compounds,[290] which was copied from an article by Wierda and Pazdernik[291] on the same subject (see below).

Alsabti, according to information gathered by Broad,[288] was a Bachelor of Medicine and Surgery from the Basra Medical College in Iraq. He had a Jordanian passport with which he arrived in the USA in 1977 on a scholarship. He first worked for a month under Dr Hermann Friedman at Temple University in Philadelphia, then at the Jefferson Medical College in Philadelphia until April 1978. In September 1978 he arrived at the laboratory of Dr Giora Mavligit at M. D. Anderson Hospital in Houston. In April 1980 he became a resident at the department of internal medicine at the University of Virginia in Roanoke. In June, when the plagiarism incidents were exposed, he was asked to leave the VA hospital, and since then his whereabouts are unknown. During his stay in the USA Alsabti obtained an MD degree from the American University of the Carribean (Montserrat). He also became a member of no less than 11 scientific societies.

Of the 13 articles published by Alsabti in 1979, five are indisputably plagiarised.[286, 287, 290, 294, 296] Some of the 13 articles were published by Alsabti as sole author, some with co-authors. Of the co-authors, Major General D. Hanania subsequently denied having had anything to do with one publication. The names of Omar Nasser Ghalib, Mohammed Hamid Salem in one paper,[290] and of K. A. Saleh and A. S. Talat in another,[288] seem to be fictitious. The address for reprints in Alsabti's papers varied from different residential addresses in both England and the USA, to the Royal Scientific Academy, Amman, Jordan, and the Albaath Specific Protein Reference Unit in Baghdad.

Why did Alsabti change his place of work in the USA so frequently? Hermann Friedman, at Temple University, Alsabti's first employer, had doubts about Alsabti's integrity when the latter claimed he had a vaccine against leukaemia, but would not give details as to what it was or how it was prepared. Wheelock, who put Alsabti on a programme of clinical cancer research, was told by some members of his team that Alsabti was making up data for his experiments, so Wheelock asked him to leave the laboratory. As was found later, Alsabti took with him a copy of Wheelock's grant application which had been written four months before Alsabti's arrival.

While in M.D. Anderson Hospital, Alsabti pirated the paper of a PhD candidate from the University of Kansas, a Dr Wierda. The latter had submitted his manuscript to the *European Journal of Cancer*. The editor had sent the manuscript for review to Dr Jeffrey Gotlieb at M.D. Anderson, not knowing that Dr Gotlieb was now dead. According to Broad,[288] Alsabti took the manuscript out of the mailbox, and used it as his own paper; he submitted it with fictitious co-authors to the *Japanese Journal of Medical Science*.[290] Alsabti's paper differs from that of Wierda and Pazdernik only in its title, the acknowledgements and the lack of references to Pazdernick's previous work. The rest is virtually identical. When Wierda saw the article, published under the name of Alsabti, in the Japanese periodical, he wrote to the editor, Akiro Shishido, expressing his suspicion that his paper had been pirated during the reviewing process for publication in the European journal. Akiro therefore wrote to Alsabti as well as the head of developmental research at the M.D. Anderson Hospital, Dr Freireich. Alsabti did not reply. Dr Freireich informed Shishido that Alsabti had never conducted any such research at M.D. Anderson Hospital and that his paper was therefore most probably a case of plagiarism. In view of this information, Shishido announced in *Nature*,[292] as well as in the *Japanese Journal of Medical Science and Biology*,[293] that Alsabti's paper was retracted.

According to *Nature*,[292] another paper by Alsabti, on fatty substances in blood of patients with cancer of the liver,[294] may also have been plagiarised, from an article by Yoshida *et al.*[295]

In another editorial *Nature* (1980, 286: 433) expressed the opinion that Alsabti's paper on lymphocytes in breast cancer [296] was virtually identical with an article published by Sylvia M. Watkins[297] in *Clinical and Experimental Immunology*. The only difference between the two papers was that in his version Alsabti added to the list of references his own paper entitled 'Lymphocyte transformation in bladder carcinoma (in press)', which in fact was never submitted nor published.

An editorial in the *British Medical Journal* (5 July 1980, p. 41) describes two additional examples of Dr Alsabti's plagiarism.

Alsabti's piracy raises the problem of how editors of scientific journals can know that a similar or identical paper has been simultaneously, or nearly simultaneously, submitted to another journal (Wierda's paper reached the *European Journal of Cancer* in October 1978, while Alsabti's paper was received in Japan in November 1978). In his letter to the *Lancet* Wheelock suggested that in order to avoid embarrassment editors should verify the credentials of individuals who have never published original research papers on the subject before. "This can be done by authenticating personal communications and acknowledgments cited in the article, and by requesting reviews of such articles by individuals who are prominently referenced.'

PILTDOWN ADDENDUM

Our consideration of plagiarism brings us back to an incident long forgotten, uncovered by Weiner[266] while he was investigating the Piltdown hoax, discussed in the previous chapter. It appears that Dawson, who according to Weiner was the main suspect in the Piltdown forgery, plagiarized an unpublished manuscript.

In 1910 Dawson published a two-volume book entitled *The History of Hastings Castle*. This book was favourably reviewed and became a standard book of reference. In 1952, Mr Mainwaring Baines, the curator of Hastings museum and a student of history of Hastings Castle, found a manuscript by one William Herbert, who carried out excavations at the castle in 1824. For unknown reasons, Herbert did not publish his work; but to Baines's surprise, half the material which Dawson had used in his book was copied from Herbert's manuscript.

It seems that this was not the only case of literary piracy in which Dawson was involved. Dawson was an ardent collector of palaeontological material and of iron objects, which he exhibited in 1903 and 1908. In 1903 Dawson published an article about iron in *Sussex Archeological Collections* (5: 46) in which 27 out of 61 pages were copied word for word from an earlier writer, Topley, to whom no acknowledgement was included.

SELF-PLAGIARISM

In addition to plagiarism, *sensu strictu*, where a scientist uses or copies the material or a paper of another scientist and publishes it under his own name, there occur also cases of self-plagiarism. In such a situation the scientist, or a group of scientists, submit the same, or almost the same article to more than one journal. Simultaneous publication of the same material in two or more journals, dual submission of manuscripts, as well as publication at short intervals of material that may be different in form but not in content, is considered to be irresponsible.[298] All journals include in their editorial instructions a sentence to the effect that: 'papers are considered [by the journal editors] on the understanding that they have not been published or are under consideration elsewhere'.

Why should scientists wish to plagiarize themselves? One possible reason is that papers authored by a number of collaborators are usually cited by the name of the first author followed by *et al*. The names of the others on the team do not appear in the citation, and the people involved feel that this detracts from their recognition as active researchers in the given field. The remedy then is that the authors take turns in appearing as the first-named author, but each paper published is only a slightly modified version of the same findings.

Another practice that may result in seeming self-plagiarism is that, because most authors like to have their papers published as soon as possible, and are dismayed by the frequently long delays in publication, they submit the same paper to two or more journals.

Once the paper is accepted for publication by one, they retract their manuscripts from the others. Nevertheless, it may happen that the retraction is not made on time, with the result that the paper is published in two journals. This happened to Mark W. J. Ferguson from Northern Ireland. He studied the microbial degradation of the outer calcified layer of an alligator eggshell and published the results of his investigations in two articles, one published in *Science*[299] and one in *Experientia*.[300] The summaries, the photographs and the figures are identical in the two papers. Later, in the June 1982 issue of *Science*, Ferguson apologized: 'I had intended to withdraw the *Experientia* paper if the manuscript for *Science* was accepted. However, due to a gross oversight on my part, I failed to withdraw it.'

The practice of repetitive publication was condemned by Lock in the *British Medical Journal* in 1984.[301]

In an interview in the *New York Times* (14 December 1982) Benjamin Levin, the editor of *Cell*, distinguished three kinds of duplication of articles. The first is the kind described in the above example. The second involves publishing a preliminary report followed later by a more detailed paper which does not materially add to the information contained in the first report. A third type of self-plagiarism consists of re-hashing results already published in primary journals in a collection of papers at a symposium. All these procedures hinder effective communication among scientists because they steal the time of both readers and reviewers, they put an additional burden on the editorial staff of journals and they clutter up libraries. Moreover, all this costs money, which is supplied by someone other than the author. What can be done to eliminate this practice?

The *New England Journal of Medicine* led the way in 1969 with the so-called 'Ingelfinger rule' named after Franz Ingelfinger one of the editors of the journal. This rule stated:

> papers are submitted with the understanding that they or their essential substance have been neither published nor submitted elsewhere (including news media and controlled circulation publications). This restriction does not apply to (a) abstracts published in connection with meetings or (b) press reports resulting from formal and public oral presentation. Editorial, *New England Journal of Medicine* 1969, 281: 676)

The *British Medical Journal* has reiterated these rules and even extended them: ' . . . the practice of publishing almost identical articles

in supplements reporting conference proceedings as well as in main
line scientifc journals must stop'.[301] In addition, the *BMJ* editors
plan to inform *Index Medicus* about infringement of these rules and
request that it print prominent statements naming the authors practising repetitive publication.

The journals *Cell* and the *Proceedings of the National Academy of Science*
now inform authors that if they publish an article which has been
submitted by them elsewhere, they will be barred from publication
in these journals for the following three years. The *New England Journal of Medicine* goes even further: it will not publish any study that
has been previously reported in the mass media, claiming that this
rule will prevent serious scientists from permitting the mass media to
present their work to the public in an inaccurate and incomplete
way.

The Publication Board of the journals of the American Society of
Microbiology (ASM) announced in its *ASM News* (1984, 50: 106)
that it would insist on joint responsibility of *all* authors of a submitted paper, and that should the paper contain plagiarized material,
be submitted simultaneously to different journals, include improper
use of personal communications, and omit citation of relevant work
by others, the following steps would be taken against the 'offender':
reprimand (for minor infractions), or a three-year prohibition
against submitting papers to any ASM journal by *any* of the joint
authors.

If indeed all main line journals would institute similiar rules and
procedures, the scourge of plagiarism and self-plagiarism would be
eliminated from the scientific literature.

With the increasing number of scientists and technicians participating in modern scientific, medical and industrial research,
there is always a problem of attribution of credit. The main author
may sometimes conveniently forget the contributions of others who
helped. This conscious or subconscious forgetfulness may be labelled
as 'quasi-plagiarism'; it has indeed been the subject of some action
taken by the International Committee of Medical Editors (representing 12 medical journals).[302] The manuscripts submitted for publication must now be accompanied by a letter including a statement that
all the authors approve the paper, as well as providing information
on prior or duplicate publications or submissions. All contributions
(intellectual as well as technical) have to be acknowledged, with the
prior consent of the persons named.

ARE EDITORIAL BOARDS INFALLIBLE?

The issue of plagiarism which we have been considering gave rise in one case to an interesting controversy over the accountability – or infallibility – of journal editorial boards. In December 1982 Dr Isaac Ginsburg of the Faculty of Dental Medicine at the Hebrew University in Jerusalem wrote a letter to the editor of the *American Society for Microbiology News*[303] complaining about the publication in an ASM journal (*Infection and Immunity*) of a paper by two authors omitting all references to the relevant work of Ginsburg dealing with enzymes present in the walls of Streptococci (eight papers and reviews published by Ginsburg during the preceding two decades). Thus the impression was created that the original findings of Ginsburg had been made by the two authors. The complaint Ginsburg lodged in the *ASM News* was that an ASM journal would not consider taking any action about the matter except recommending a private apology by the two authors to Ginsburg. Ginsburg carefully refrained in his letter from alluding to the fact that one of the authors was a member of the editorial board of *Infection and Immunity*.

The editor of *Infection and Immunity*, and the chairman of the Publication Board of the ASM, in their correspondence with Ginsburg, agreed that the problem of inadequate credit was getting worse, but also stated that nothing could be done about it. The editor of *Infection and Immunity* wrote:

> I realize that many authors fail to acknowledge prior work. It is now so commonplace that I know of few investigators who have not been slighted this way . . . I prefer to consider the author's apology as the end of this particular incident.

Another ASM official to whom Ginsburg addressed his grievance evaded the issue by stating:

> I am unable to intervene in any manner because your letter to me is headed by the words 'Confidential and Personal', that in itself prevents me from even trying to intervene.

I find this a very 'elegant' way of evading the issue!

In his letter to the *ASM News*, Ginsburg stated that the authors, whom he had contacted about the matter, had admitted their misdemeanour and had apologized; they had also claimed that the omission of the key literature was an oversight. They refused, however, to correct this oversight in public.

The chairman of the Publication Board had evaded the issue by claiming that the ASM journals had no mechanism for publication of such apologies and called the attention of the readers to the decision of the Journals Subcommittee refusing requests made by T. D. Mukhur of Glebe, Australia, and by Ginsburg to permit publication of critical comments on published papers. ' . . . because of the concern that the letters would be devoted to criticism concerning inadequate citation of the literature' (Report of the Journal Subcommittee, *ASM News*, 1982, 48: 117).

Ginsburg's bitter conclusion in his letter to the *ASM News* was that 'the abused and cheated authors are asked by all parties to accept private apologies, which are not worth the paper on which they are written'[303].

Ginsburg's letter evoked a considerable response from microbiologists such as Bernard D. Davis (Harvard), K. D. Stottmeier (Boston City Hospital), Norton S. Teichman (University of Pennsylvania), Jiann Shin Chen (Virginia Polytechnic Institute) and Elvin A. Kabat (Columbia University), all demanding that the ASM Publication Board come up with some mechanism that would permit the publication of charges and rebuttals in the ASM journals.[304] It was Teichman's view that 'Such a section could be a force for disseminating timely and positive comments, but it could also act as a deterrent to those who consciously choose to compromise their integrity'. Jiann Shin Chen stressed the neglected responsibility of referees in this matter, pointing out that it is the duty of the referee to check the correctness of the information presented in the manuscript, and this includes the correct citation of previous work in the field. Even when the referee detects small errors in the submission, he should examine the manuscript very carefully to 'ascertain the soundness of the results and the validity of the conclusions'. Kabat felt that the private apology solution was absurd and that 'the ASM journals are adopting a 1984 Orwellian approach to history before its predicted time'.

In response to all the criticism, Dr Shands, the editor of *Infection and Immunity*, stated[305] that there was an error of omission without malice and dishonesty on the part of the two authors mentioned by Ginsburg and that the private apology should have been sufficient to close the matter; Ginsburg's demand for a public apology, published as a letter to the editor, was not feasible, he said: first, ASM journals did not publish letters to the editor, and secondly an editor could not require of an author that he publish a public apology. This statement is based, of course, on the earlier decision of the Journals Sub-committee not to have a section for letters to the editors in ASM journals.

Dr Helen R. Whiteley, chairperson of the Publication Board, stated that 'publication of unvalidated accusations would not be fair' and that 'the validation of complaints would require extensive investigation'. She therefore was of the opinion that it was the task and responsibility of the referees (three for each paper) to monitor the proper citation of the literature pertaining to the manuscript. She then suggested establishing an Author's Corrections section in the ASM journals which would permit authors, in case of a complaint, to correct published work and to cite omitted references. This would be a workable idea, however, only if both the complaining party and the original author agreed that an error of omission had indeed occurred. Lack of agreement would leave the matter unresolved and the aggrieved scientist would have no recourse.

What is the scale of the problem we are considering here? During 1982 all the ASM journals taken together published 7000 articles, and there were only two complaints. Since these journals have in total about 750 editorial board members, the yearly average load for each member of the board is at least ten articles (and probably double or triple that figure if one considers that some of the submitted manuscripts are rejected). Furthermore, the board member has to find three referees for each paper submitted. It is assumed that the editor chooses as referees the top people in the given discipline, whose expertise is such that they would be familiar with all the relevant literature and would thus easily discover any errors of omission. In fact, we know that the recognized authorities in any subject are burdened not only with their own research and directing the research of their collaborators and students, but also by many administrative duties in universities or research institutions; moreover, they are in great demand to appear at seminars, conferences and workshops all over the world. The solution to the additional demand to act as referees is in some cases solved by delegating the task to junior staff members (or even PhD students working on a similar problem), and here lies the danger that an omission would not be discovered and the paper passed for publication because its data were acceptable and of importance.

Because of this consideration, passing the buck to the referees, though theoretically correct and absolving the editorial boards from some of their responsibility in relation to 'slighted' authors, is obviously not the solution. The only promising suggestion is that made by Dr Whiteley, i.e. to have a section in the journals where such matters could be aired in an appropriate manner.

A further problem concerning fair handling of manuscripts by editorial boards has been raised by B. Max, writing in 1984. Writing about the Darsee affair,[222] Max observes that the *New England Journal of Medicine* appears to have bias towards Harvard and Boston. In 1981, out of 146 original articles published in the journal, 23 were authored by scientists from Harvard and eight more were from the Boston area (22 per cent). Though the name of the journal implies that it is a regional publication, it is actually considered to be one of the important and influential medical journals in the world. Nevertheless, only 21 per cent of articles came from outside of the USA. Max believed that all submissions should receive equally rigorous reviews, and that editorial decisions should not be mollified by personal knowledge of the author or his institution. A fault in Max's arguments may lie, however, in the possibility that significantly fewer articles were submitted from outside the Boston area.

LITERARY FRAUDS – REVERSE PLAGIARISM

We encounter plagiarism in science when one scientist copies the results of a paper by another, and publishes them under his own name. The literary world, however, is not free of its own villains. There exists here another mode of unethical behaviour – the fabrication of non-existent work under the name of a famous author or person. Such literary forgeries have been occurring since antiquity.

A document called the *Constitutium Constantini* was allegedly composed in the eighth century. In this document the emperor Constantine ceded to Pope Sylvester and his followers the rule of Rome, Italy and all the countries to the west. The document remained hidden for some 200 years, and was then first used by the Pope of that period as a proof of legitimacy of papal rule over Italy. In 1440 the scholar Lorenzo Valla exposed the document as a forgery in an essay 'De falso credito et ementitate Constantini donationi declamatio', later published in 1517. Vella based his findings on an analysis of the character of the Latin language of the time when the document had been allegedly composed. The authenticity of the document nevertheless continued to be a subject of polemic until the end of the eighteenth century.

There were at least two known literary frauds in the eighteenth century. In one, the 12-year-old Thomas Chatterton wrote in 1760 some poems which were then circulated as if written by a fifteenth century monk, Thomas Rowley. The fabrication was exposed some five years later, and contributed to Chatterton's suicide.

Similarly, in 1763 a Scotsman, James MacPherson, published an epic in six books entitled 'Fingal', as if written by a third century bard, Ossian. The original poems were supposed to have been written in Gaelic, and MacPherson made it known that he had translated them, when in fact he was their sole originator.[306] Though his contemporary Samuel Johnson denounced the Ossian poems as fakes, and MacPherson refused to display the original manuscript, 'Fingal' enjoyed great popularity: Goethe praised it and Napoleon decorated his study with paintings of scenes from Ossian. The proof of forgery came long after MacPherson's death in 1790. (He was buried in Westminster Abbey.)

This century has witnessed at least four great literary forgeries.[306] The first occurred in 1928 when the *Atlantic Monthly* published a series of articles on 'Lincoln the lover', based on an exchange of letters between Lincoln and Ann Ruthledge. The letters reached the journal via a former actress and columnist from *San Diego Union*, Wilma Frances Sedgwick, who claimed the letters had been handed down through the family of her mother, Cora de Boyer. The letters were later shown to be a forgery perpetrated by Cora de Boyer herself.

The second forgery took place in 1947. Rosa Panvini and her daughter, from Vercelli in Italy, sold to the newspaper *Corriere della Sera* letters allegedly written by Benito Mussolini. The authenticity of these letters was vouched for by Mussolini's son, Vittorio, as well as by an expert from Switzerland. The two women said the letters had reached their father and husband from a minister of Mussolini near the end of the war, with the request that they should be hidden. There was also a set of 30 volumes of diaries. The Italian police later charged the women with forgery and fraud and confiscated 26 of the volumes. The women were convicted and given a suspended sentence. Nevertheless, they managed to sell the remaining four volumes to *The Sunday Times* for a considerable five-figure sum. The newspaper learned about the hoax too late to recover the money, but managed to abort publication of the fakes.

In 1971 came Clifford Irving's alleged 'autobiography' of the reclusive millionaire Howard Hughes. Irving forged letters from Hughes to himself, as well as a contract between the two men authorizing Irving to write the Hughes' biography. Irving used these documents to convince McGraw–Hill publishers that he was authorized to write Hughes's 'autobiography', and this resulted in a 750,000 dollar contract between McGraw–Hill and Irving. After a number of coincidences and involvement of a previous associate of Hughes, Noah Dubrich, as well as some detective work, Hughes himself denounced Irving's work as a hoax. At that stage Irving confessed his misdeeds and was sentenced (together with his wife and accomplice) to 16 months' imprisonment.

This series of frauds over the past 50 years was topped by the forgery of Hitler's diaries.[307]

On 22 April 1983 the editors of the German weekly *Stern* announced the discovery of Hitler's diaries – 62 volumes that covered the period from 1932 until 1945. *The Sunday Times*, *Paris Match* and the Italian *Panorama* bought the rights to publish these diaries for huge sums of money. According to the editors of *Stern* the diaries were purchased from an undisclosed source by their correspondent Gerd Heidemann, a great Nazi sympathizer. According to Heidemann, a week before Hitler's death in Berlin, he despatched all his precious possessions and the diaries in a plane which crashed near Dresden in April 1945. Unidentified persons had recovered the diaries from the wreckage and hidden them.

The announcement in *Stern* became one of the publishing sensations of the century, especially after the authenticity of the find was vouched for by Cambridge historian Hugh Trevor Roper, an expert on Hitler. He was the author of the book '*The Last Days of Hitler*' and was familiar with Hitler's handwriting. There were, however, several scientists and handwriting specialists who pointed out that Hitler had suffered a palsy which had worsened considerably after the attempt on his life in 1944 and which should have affected his handwriting; moreover, Hitler's custom was to dictate to his secretaries and none of his contemporaries who were still alive, knew that Hitler had kept diaries.

Charles Hamilton, an autograph dealer in New York, saw photocopies of the alleged Hitler diaries and pronounced them not to be credible; so too did historians Marie Barnard, Werner Maser and Joachim Fest.

As the voices of doubt increased, Trevor Roper reconsidered his opinion and finally admitted that he might have been wrong, and that indeed some documents in the collection could be forgeries. Nevertheless, Peter Koch, the editor in chief of *Stern*, flew with a number of volumes of the diaries to New York to display them in public and to defend them as genuine. A handwriting analyst from Newton, Massachusetts, Kenneth Rendel, to whom Koch showed the diaries, photocopied and enlarged some of the pages and concluded they were forgeries. In spite of this, *Stern* went ahead with publication of the first instalment of the diaries, including a segment purporting to show that Hitler knew and approved of the flight of his deputy Rudolf Hess to England in 1941.

At a press conference after the publication of this part of the diaries, a chemical expert, Luis Ferdinand Werner, examined the diaries and found that the paper, the cover and the binding, as well as the glue and the labels, originated from the post-war era (the binding, for instance, contained polyester fibres, which did not exist before the war). He concluded that the diaries were fakes. It was also discovered that the text of the diaries was a plagiarized edition of a book of the Federal Archivist of the Nazi regime, Max Domarus, the compiler of 'Hitler's Speeches and Proclamations 1932–1945'. The diaries contained a number of errors identical to those made by Domarus (such as the statement that the crowd at a Breslau rally numbered half a million, when in fact more reliable reports of the event estimated it at 100,000).

Following this revelation, two editors of *Stern*, Koch and Felix Schmidt, resigned. The *Daily Express* in London wrote: 'It was the day the thunder of *The Times* turned into a whimper and *The Sunday Times* was forced to sniff the stench of self-deceit'.

Heidemann, who was thoroughly questioned about the source of the diaries, did not budge from his story, but he was forced to resign from *Stern*.

In the court case which resulted, the truth came out. The real forger was Konrad Kujau, a dealer of Nazi memorabilia who had duped Heidemann, described by the historian Maser as a 'gullible person morbidly interested in Nazi paraphernalia', into believing in the authenticity of the diaries. The court case against the two ended in June 1985 in Hamburg. Judge Hans Ulrich Schroeder sentenced Kujau to four and a half years imprisonment and Heidemann to four years and eight months. Half the sum of three million dollars paid over by *Stern* had vanished (*Newsweek*, 22 July 1985, p. 22).

Writing in 1983 about the forgery of Hitler's diaries, Ed Magnuson of *Time Magazine* criticized *Stern* for placing 'journalistic

expediency above society's overriding need for accurate history.
There is never a need, nor a justification for publishing first and
authenticating later . . .'.[307]

11

How Honest Are Grant Applications?

In the early days of modern science, that is, science as we know it to-day, this field of human activity and its practitioners enjoyed public prestige. Scientists were respected as enlightened representatives of the quest for knowledge. In those early days people believed that all scientific discoveries benefitted mankind. With the passage of time, many results of scientific investigations were applied to the military and technological fields. This turn of events meant that the funding of science became more and more dependent on social and political pressures. The patronage by benefactors or public funds came to be replaced by government funding. This situation had a downgrading effect on basic science and research, and began to erode the ethical standards of behaviour among scientists.

In developed countries most of the financing of scientific research comes either from government or from industry. Some research is also supported by private foundations, and lately even by the stock market. In the USA, for example, Congress determines how much money is allotted to the National Institutes of Health (NIH), and there are 11 such Institutes. The approval of the budget depends on the lobbying activities of public interest groups and foundations. This system invites pressures on the scientists to direct their research efforts into areas they perceive to be well funded, rather than the areas they might genuinely be most interested in. During the Nixon administration, for example vast resources (about 100 million dollars) were earmarked for cancer research; any scientist who could indicate that his research even marginally connected with cancer had a very good chance of being funded.

Since one of the factors taken into consideration by study sections of the granting agencies is the status of the applying investigator and his recognition as an authority in the specific field of science, scientists are well aware that a brilliant idea *per se* is not sufficient to serve as a basis for a grant application and to have it approved. The applicant has to demonstrate not only that he is well versed in the field, but also that he has already made some significant contributions to it. This is documented by the number and the quality of his published papers. There is a widespread belief that it is usually (but not always) the number of papers that carries most of the weight in the judgement of the relative merits of the applicants, be it for their promotion at the universities or research institutions, or for their grant applications. The phrase 'publish or perish' is the quintessence of this belief.

An important outcome of this pressure to publish (especially when research funds are scarce and the competition strong) is that it provides an incentive for cutting corners and for scientific fraud. The latter may assume many forms, ranging from dishonest structuring of the grant application, via plagiarism and fudging the supporting data, down to outright falsification of the research results.[308] Dr Edward Huth, editor of *Annals of Internal Medicine*, has classified the possible abuses in this field as 'false authorship', when the department chairman or a laboratory technician, not responsible for the intellectual content of the paper is included as one of its authors; or 'salami science', i.e. repetitive publication of the same material.[309]

Nevertheless, the investigation of frauds by NIH revealed that out of some 20,000 projects handled by NIH each day, there are hardly two reports per month that may be considered as verging on scientific misconduct.

Greenberg[310] stated:

A commonly offered explanation for the alleged increase of incidence in fraud is that the present scarcity economy puts researchers under unprecedented pressure to make their mark.

This argument is debatable. Most cases of fraud have been discovered in institutes and universities, and there has not been a drastic reduction of federal grants given to scientists in these institutions. Second, it seems that the record of cooking laboratory books is not worse for big research teams in comparison with small ones or with individual researchers. Thus the pressure to publish and the scarcity of funds may be considered as only contributing factors, not the mainspring of scientific misconduct.

Leigh van Valen, Professor of Biology at the University of Chicago, claims that an honest grant application is incompatible with conceptually original research.[311] Exciting work by honest people seldom gets funded, he says. The problem is that once a researcher has committed himself to the approved original proposal, he cannot appreciably deviate from his research plan. The usual duration of research grants is three years. If during that time the scientist develops some new ideas, or original approaches, he can pursue them only at the cost of not satisfying the demands of the grant proposal. Indeed, such a situation occurred in the case of Jim Watson's celebrated work on the structure of DNA with Sir Francis Crick; a strict adherence to the approved programme for the post-doctoral fellowship which Watson held from the National Science Foundation would not have permitted him to carry out the research that won him the Nobel prize.

How then do scientists ensure that they are able to do good and innovative research when they are applying for a grant?

According to van Valen[311] there are several ploys that can be used. First, the applicant can describe in the grant application work he has already done, but which is not yet published. In this case the applicant already knows the results, and can therefore describe the proposed experiments, knowing that they will succeed. Once such a grant application is approved, the applicant can use the funds for new original research which was not mentioned in the grant application at all. This is, of course, basically a dishonest procedure but I believe it to be quite often practised.

Another possibility is to apply for a grant in a field where any result, positive or negative, will be of importance. This involves testing an established hypothesis, searching and screening for drugs, their clinical evaluation, discovering whether or not a known phenomenon applies to a new particular case etc. Such research is intellectually less exciting but has a better chance of being funded than research based on entirely novel ideas which may well fizzle out completely. Thus creative science and new ideas seem to have less chance of being supported, unless they appear to serve the national or the industrial interest in some way.

THE ROLE OF THE PRINCIPAL INVESTIGATOR

Since most research funding depends on the principal investigator (PI) of the project, his or her role in the ethical execution of the research has to be assessed. The PI is defined not only as the person applying for, and receiving the grant, but also as the person who assumes total responsibility for the execution of the project, and in the end submits a paper for publication (or writes the report) and puts his or her name among the list of authors.

How can it be ensured that all published data are true and reliable? Some recent frauds and forgeries have involved large research groups, often spread over geographically discrete institutions, where the PI or the head of the laboratory did not directly participate in, or personally supervise, the conduct of the experiments (for example, the case of Braunwald at Harvard, or Felig at Yale).

In a large group headed by a PI there may be graduate students, or 'post-docs', or clinicians each having an expertise in the use of a particular technique or of an instrument, at which the PI himself has no first-hand experience. Consequently, when results based on this technique or instrument are presented to the PI for review or discussion, he actually has no real means of verifying the reliability of the results and has to rely on the honesty of his collaborators. This honesty may, however, be hard pressed, especially in a situation where a grant or project is due to come up for review and there is an urgent need to produce data. In such a situation there is an incentive not necessarily to produce fraudulent data, but to select from the array of conflicting results those that best fit the idea propounded in the grant application. So, how important is it if one omits from the graph some points that would wreak havoc with the proper course of a curve; or when one combines the numbers from different experiments in which the controls are differently set; or conveniently forgets all sorts of mishaps that invariably occur in any experiment; or when one use the results of just one single experiment to support a given contention?

In any case 'the further the senior PI is from the day-to-day activities of the laboratory, the more likely is it that fraud by a subordinate can be perpetrated'.[312] If the PI accepts exciting results without carefully checking them, and rewards the person who produced them, and on the other hand puts pressure on another who does not come up with the expected results, the ground is being prepared for fraudulent work. In such a situation at best unreliable data are being produced and circulated and, at worst, a fraud is perpetrated.

It is therefore the responsibility of the PI to go over the *raw* experimental data, as well as to check the process of converting these data into publishable results. This involves checks on actual counts (be it colonies of bacteria, plaques of viruses, animals, radioactive counts or instrument readouts, or micrographs, as well as on statistical and graphical methods of obtaining averages and interpolated values). One important safeguard is that each experiment be repeated in its entirety at least once, and if possible several times, by another member of the research group.

One of the most touchy points in any experiment concerns the controls. Every experiment needs a negative control, i.e. where one omits the essential ingredient that is being tested and obtains a negative result. Some experiments also demand a positive control. For instance, when testing for the presence of a quantity of a certain drug in body fluids, a pure drug (known to be active) is included in the test to ensure that the whole experimental set-up works properly and is able to detect the presence of the drug. If in such an experiment the positive control gives a negative result, the finding that no drug is detected in the clinical test sample is meaningless.

Often, some crucial controls are omitted inadvertently, and sometimes with the intention of saving the labour of setting them up. In experiments that involve the routine repetition of procedures some controls are omitted because 'they always come out the same way', so why repeat them in each and every experiment? The famous French saying 'Cherchez la femme' may be converted in science to 'Cherchez le contrôle'.

Another of the responsibilities of the PI is to see that the work is performed in a careful and orderly manner; sloppiness and negligence should be discouraged and prevented. Finally, the PI must write up the results and send the paper for publication. The PI should never allow his (or her) name to appear on a paper unless he is convinced of the reliability of the data and conclusions, and unless he is willing to accept responsibility for the entire paper, that is, even for those parts in the execution of which he was not personally involved. Those who do not recognize the limits of their capabilities as researchers and administrators of research may get burned, and some actually do.

The mode of action described above applies not only to principal investigators of grants, but to all scientists heading a research group, a laboratory or an institute. Their absolute trust in their subordinates does not relieve them from their obligation to participate in and monitor all the research activities within their area of responsibility.

12

Other Ethical Problems in Science

AN AUTHOR CONTRADICTS HIMSELF

I have discussed in detail in chapter 10 the various forms of plagiarism. Unethical behaviour of quite the opposite nature is this case of presenting misleading information by authors failing to cite their own identical research work giving contrary results.

A group of scientists from the National Institute of Mental Health who studied the presence of an active enzyme, monoamine oxidase (MAO), in paranoid schizophrenics, published a paper in the *New England Journal of Medicine* in January 1978 stating that MAO activity in small blood elements responsible for the clotting of blood was significantly lower in chronic schizophrenics than in normal controls.[313] (MAO degrades norepinephrine, a natural hormone of the autonomous nervous system which raises blood pressure and accelerates the heart rate.) The paper evoked considerable interest and correspondence in the pages of the journal.

One of the correspondents, Dr Karen Pajari[314] was puzzled that although Potkin and his associates had found lower activity of platelet MAO in chronic paranoid schizophrenics, there appeared almost simultaneously an article in the *American Journal of Psychiatry*[315] stating that there was no difference in MAO activity between two comparably diagnosed groups; what puzzled Dr Pajari was that two of the co-authors of the paper in the psychiatry journal, Drs Murphy and Wyatt, were also co-authors of the Potkin paper. Therefore, these two authors had published, almost simultaneously, seemingly contradictory results. Other correspondents (Baldessarini, Lipinski) commented that the level of MAO was indeed controversial, since there had been some ten reports published earlier

that confirmed the decreased MAO activity in chronic schizophrenia, while nine other papers reported a failure to find any
difference between MAO activity in schizophrenic patients and normal controls.

Potkin, Cannon, Murphy and Wyatt responded to the remarks
made in the correspondence and stated:

> Dr Pajari notes that two of our authors recently presented a platelet
> monoamine oxidase study that did not demonstrate subtype dif-
> ferences within schizophrenic patients. We could not previously
> address ourselves to the then unpublished study by Berger et al.
> because it has been our policy not to discuss unpublished data in a
> published paper.[316]

In a summary of the correspondence, the editor of the *New
England Journal of Medicine*, Arnold Relman, stated that the editorial
board was indeed puzzled by the simultaneous publication of two
apparently contradictory papers which shared two authors:

> . . . neither manuscript, as submitted [in March 1977 to the *American
> Journal of Psychiatry* and in July 1977 to the *New England Journal of
> Medicine*] referred to the existence of the other, and neither we nor the
> editors of the *American Journal of Psychiatry* (personal communication)
> were told of the existence of the other work . . . To dismiss one's own
> discrepant results as being 'unpublished data' and therefore not open
> to comment defies common sense and is, to say the least, dis-
> ingenuous.[317]

The epilogue to this story was provided by a letter to the editor of
the *New England Journal of Medicine* signed R. J. Wyatt and D. L.
Murphy:

> We sincerely regret not having pointed out our discrepant results
> from a second study in our recent article on platelet monoamine ox-
> idase and schizophrenia . . . [318]

Writing about this incident in *Science*, Broad[319] wonders who
actually takes the responsibility in multi-authored papers. The
multiple publication of the same data, accompanied by a decreasing
length of the paper, co-authored by a bevy of researchers with slight
changes of the composition of the author cocktail, not only con-
tributes to an exponential growth of literature, and to padding of
bibliographies in CVs but it also leads to publication of premature
studies and causes a serious loss of time among scientists who are
trying to keep up with the literature.

It may also happen that an author included in a multi-author paper may not even know that he had been included. George H. Arronet, who found his name among 12 others as an author of a paper on amenorrheoa and oral contraceptives in *Fertility and Sterility* September 1979, wrote to the editors disclaiming his participation in the article and stating that he actually disagreed with the conclusions drawn from the study.[319] The lead author admitted that an error had occurred.

ETHICS IN THE POPULARIZATION OF SCIENCE

In her book *Reflections on Science and the Media* June Goodfield[180] discusses the problem of media coverage of scientific research. She presents the strong and the weak points of science journalism as it had handled the DNA recombinant controversy, Summerlin's painted mouse, the publication of Rorvik's book *In his Image* and the thalidomide cover-up in England. Let us turn our attention to two of these incidents in particular.

The DNA recombinant problem arose when it became technically possible to transfer pieces of DNA (or genes) with the aid of plasmids (independent replicating DNA structures that exist in most bacteria) from one bacterium to another or even from an animal cell to a bacterium. The scientists then showed that the bacteria containing such added genetic information could translate it into proteins. At the famous meeting at Asilomar in 1975 scientists expressed their apprehension that this new technique of genetic engineering would result in the creation of new bacteria that might cause incurable diseases of animals and plants.

The scientific discussion on this problem was taken up by the mass media and this resulted in a greatly exaggerated public scare and fear of the recombinant DNA. It took several years for the agitation to subside. We now know that research in this new field has actually contributed to improvements in the well-being of humans and animals and is on its way to revolutionizing agriculture also. Insulin, interferon, and growth factors are just a few examples of recombinant DNA products, produced now on an industrial scale. The public outcry raised by the media did much to slow down progress in this field and also undermined to a certain degree people's faith in the progress of science.

Did scientists themselves contribute to this situation? Goodfield[180] finds that scientists suffer what she calls 'mental inbreeding', leading them to seek contacts with other members of the professional community, but to avoid interaction with the media under the pretext that they distort scientific facts in their popular descriptions.

Scientists who seek acclaim and recognition among their peers have an ambivalent, if not antagonistic, attitude towards anyone who popularizes science. Such a scientist is thought to demean himself. This attitude leads to a 'loss of the sense of personal and professional responsibility for the influence of their [scientists'] work on society. Because the allegiance is to be scientific disciplines and not to the society, moral responsibility is relegated to the background.'[180]

Scientists also distrust the popularization of science because they do not think it proper for a scientist to seek more public exposure than his work deserves. The real danger of popularization comes when a scientist who inflates himself as an expert offers data and speculations without making distinctions between them, and encounters a gullible or perhaps a careless reporter.

It is, however, also possible that the scientist's dislike of popularization is dictated by jealousy of the success of fellow scientists. I still remember the scathing criticism received by Watson's book *Double Helix*, mostly for its popular but also subjective and personal account of the events that had led to the discovery of the structure of DNA, and eventually to the Nobel prize.

With the increasing fragmentation of science and the ever more rapid proliferation of specialized jargon, understandable to only a few members of the 'invisible college', there is a need for popularization among the rank and file of scientists in all disciplines. In addition to magazines like *Scientific American*, *Discovery*, *Science Digest*, *The Sciences*, *New Scientist*, *Omni* (to mention a few) which are catering for scientists as well as the educated public, the well-established periodicals like *Science* and *Nature* now have sections entitled 'News and Comments' or 'News and Views', that illuminate in accessible language the developments in various fields of science. Scientists are thus able to follow the important events in fields outside their own expertise.

The controversy over genetic engineering, and the analysis of developments leading to the thalidomide tragedy (see chapter 13, pp. 187–9), raised serious questions about the motivation of scientists. The image of scientists as disinterested seekers of truth and as public servants has been questioned. In these instances scientists reacted to public criticism by claiming superior knowledge of the intricacies involved, to which the public or the press have no access. The journalistic treatment of the genetic engineering problem, though initially raised by the scientists themselves, revolved about purely imaginary dangers which in the end turned out to be non-existent. The result of the scare raised by the media, however, was the imposition of restrictions and laws which for a while brought the research in this field practically to a standstill.

In 1979 David M. Rorvik published, through J. B. Lippincott of Philadelphia, a highly controversial book entitled *In his Image: The Cloning of Man*. In this book he described, as factual, a case of an eccentric millionaire, known under the name of 'Max', who had arranged for himself to be cloned so as to produce a perfect copy of himself. Rorvik related the story as if he were personally involved. He introduced to Max a gynaecologist, who set up a laboratory on a tropical plantation belonging to Max, and there implanted nuclei of Max's cells into enucleated eggs obtained from hired women; the eggs were then reimplanted into the wombs of the women. (For discussion of cloning see also chapter 8, p.124.) In December 1976 a child was born to one of the surrogate mothers.[320]

Most scientists ignored the book, but soon after it was published a British geneticist from Oxford, Dr D. Bromhall, filed a law suit against both Rorvik and Lippincott on the grounds of fraud, deceit, invasion of privacy and libel.[321, 322] Rorvik had mentioned Bromhall in his book as the scientist who had actually developed cloning techniques in his experiments on rabbits. It was these techniques that had been used in the cloning of Max. Bromhall thus found himself unwittingly vouching for the accuracy and the credibility of the book, while in fact Rorvik had merely written to Bromhall asking for a description of the current state of his cloning experiments; the letter was written five months before the alleged birth of the cloned baby.

After three years of litigation, Judge John P. Fullham, of the US District Court in Philadelphia, ruled that the book was a work of fiction, that the cloning described in the book never took place and that 'all the characters mentioned in the book, other than the defendant Rorvik, have and had no real existence'.[323]

At the Annual Meeting of the Society for Social Studies of Science, held in Philadelphia (October, 1982), Rae Goodell discussed the book in terms of the roles played in its acceptance by the publishing community, the mass media and scientists. She accused the publishing community of professional negligence: Lippincott did not check the manuscript for a possible fraud, and did not seek advice from scientists, or science writers, nor even from their own medical textbook experts. The only person asked for advice was a science fiction writer. The publisher included in the book a disclaimer: 'The author assures us it is true. We do not know'. Other publishers did not express any concern about Rorvik's book, although some thought that Lippincott should have asked for professional advice concerning the veracity of the manuscript.

The press coverage of the book and subsequent litigation was based mostly on speculation rather than on facts. Rorvik was given credibility by the publication of his own statement (people tend to believe the printed word), even after he had been discredited by the court. The press published very little about the outcome of the case *Bromhall* v. *Rorvik*, nor about the court's decision that the author and the publishers write a letter of apology to Bromhall and that Lippincott admit that the book was a fraud.

Within the scientific community some biologists thought the book to be a threat to genetic research, but the effectiveness of scientists such as James D. Watson and Clifford Grobstein who tried to discredit the book as a scientific report, was low because of public mistrust and suspicion that the specific community was trying to manipulate the information and bar it from the public.

Rorvik's book, as other popular scientific frauds, generated misunderstanding. It is the publicity afforded to these frauds that adds to the general distortion and misrepresentation in the popular portrayal of science.

WHO OWNS INTELLECTUAL PROPERTY?

One issue that is essentially one of ethics, although it too offers scope for deliberate misconduct, is that of who should be the owner of the results of research. Who should be the owner, for example, of cells derived from a human body and cultivated outside it? Should it be the person who freely permitted scientists to remove the cells from his or her body? Should it be the next of kin, in cases where the donor has died? Should it be the scientist(s) who manipulated the cells so as to make them grow indefinitely outside the body, converting them into clones? Or should it be the institution in which the work on establishment of the cell lines (clones) was carried out?

The answer to these questions is not easy. When cells are isolated from the human body with the purpose of establishing from them a permanent line, special procedures have to be applied to them. For example, they may be fused with antibody producing cells. The resulting 'hybridomas' (for the discovery of which Cesar Milstein and George Kohler, working at Cambridge, England, received the Nobel prize in 1984) produce exceedingly specific antibodies with many potential applications in medical treatment, diagnosis and research. Such hybridoma lines may be used for the isolation of a particular gene which, when transferred to bacteria, will enable the production of some commercially useful gene product (such as hormones, interferons etc).

In all or some of these procedures, should the owners of the cell lines be the biotechnologists who were active in their development?

In one case[324] a Japanese post-doctoral fellow, Dr Hideaki Hagiwara, a graduate in biochemistry from Osaka University, came to the University of California at San Diego in 1981 to work with Drs Gordon Sato and Ivor Royston. His research work involved fusion of human cell line UC 729-6 (developed at San Diego) with lymphocytes from a cancer patient. The patient was Hagiwara's mother, in Japan, who was suffering from cervical cancer. Hagiwara assumed that by producing a hybridoma with his mother's cells he would be able to obtain a cell line continuously producing antibodies directed against the specific cancer of his mother. Indeed, the fused hybrid cell line which Hagiwara obtained with the aid of Drs M. Glassy and H. Handley produced monoclonal antibodies which reacted with cancer cells of various origin, but not with normal cells.

Hagiwara carried the hybridoma cells back to Japan. There, with the help of his father, a physician and owner of the Hagiwara Institute of Health, he established a culture of these hybridoma cells which produced antibodies. Hagiwara injected these antibodies into his mother, after first testing them on himself, his father and three other volunteers. Unfortunately the treatment was too late in the disease and his mother died.

When these facts became known to Royston, he became concerned that the University might have lost the chance to patent the hybridoma. Another potential loser might have been Hybritech, a biotechnology company which had funded the hybridoma project of Hagiwara, and of whom Royston was a shareholder. The embarrassment was eventually solved by an agreement between Hagiwara, Royston and the university: the University of California was assigned the patent rights to the cell line and Hagiwara was granted an exclusive licence for distribution in Japan and the rest of Asia.

The question remains open, however, whether ownership of a cell can be based on familial ties, or whether hybridomas, which are new biological entities, should belong to the researcher who developed them. Gordon Sato, in whose laboratory Hagiwara worked for some time, believes that the cell donors should be given a share of any profit accruing from the use of the cells.

The problem of cell ownership and patent rights was also the subject of a lawsuit filed by the Roche Company for a judge to determine whether Roche, in removing some genetic information from cells of human origin, was infringing upon a patentable material (cells).

The story developed in 1977 when some cells were removed from the bone marrow of a patient dying of blood cancer at a Los Angeles hospital. Philip Koeffler and David Golde of the School of Medicine of the University of California established a line from these cells, and named it KG-1. Golde then sent these cells to his friend Robert Gallo at the National Cancer Institute. Gallo tested the cells for the possible presence of an endogenous virus and discovered that the cells produced interferon, a natural antiviral substance manufactured in cells infected by viruses. With the presumed consent of Golde, Gallo gave the cells to Sidney Pestka at the Roche Institute of Molecular Biology in New Jersey. Pestka, who was testing various cell lines for the production of interferon, found that KG-1 cells were actually superproducers of interferon. Roche then contracted with C. Todd of City of Hope Medical Center to determine the molecular structure of the interferon produced by the KG-1 cells. Roche also contracted with Genentech to synthesize a piece of DNA that would be a copy of the DNA in the gene coding for the interferon molecule. The Genentech scientists isolated first from the KG-1 cells messenger-RNA coding for interferon. The DNA copied from the specific interferon messenger was then introduced into bacteria, and their clone produced large amounts of interferon.

When it became clear that the gene isolated from the KG-1 cells could be used in its cloned form for the commercial production of interferon, the question arose as to whom the KG-1 cells belonged to, as well as to who should be the owner of interferon produced by the gene isolated from these cells. Should the ownership be assigned to the patient from whom these cells had originally been taken (or perhaps to his heirs), or to the researchers who had established the line from these primary cells, or to the investigators who isolated the gene?

According to Nicholas Wade:

> . . . a researcher is expected to make any special material he has developed fully available to colleagues. In return . . . his colleagues will treat the material like property borrowed from a friend . . . , as something not to be passed to third parties or used for private gain without specific permission from the owner.[325]

The problem of ethics in this case hinges not only on the question whether Golde authorized Gallo to pass the cells to Pestka, but also on the fact that the handling of the cells by Roche and Gnentech was a secret process allowing Roche to benefit at the expense of others. The question arose should genetic information extracted from a cell be considered as belonging to the same category as a scientific theory or a computer program?

In a wider context, there is also the unresolved question as to whether there should be proprietary rights over research in a university or in an industrial laboratory. What can or cannot be patented? What is intellectual property and when and where does it conflict with public interest? The consensus is that except in the case of classified military research, scientists may assume that data from research funded by grants belong to the researchers, unless it is specified otherwise in the grant contract.

The subject is relatively new, ambivalent and full of contradictions. The dispute over the control and ownership of research will yet continue.[326] In my view, scientific research, ideas, discoveries and inventions should basically become public property. Scientific achievements, however, differ from literary and art creations (music, sculpture, theatre, cinema, TV etc). The monetary rewards for the work of authors, artists and producers are their bread and butter and essential for their survival and well-being. Except for government or community supported artists, most depend on their output for their income from sales, patents, royalties etc.

Academic scientists, however, are mostly salaried workers whose livelihood does not entirely depend (though it may vary with their successful output) on their inventiveness and productivity. The main reward for their creativity and inventions is the recognition they receive from the international scientific community. On the other hand, a good case could be made for the patenting of discoveries by the institution which is footing the bill for the research. The cost of carrying out scientific research, already formidable, is escalating exponentially. Institutions may, with justice, claim that patent rights would potentially compensate them for their enormous investment in science.

In the field of genetic engineering, the Genbank (USA) and the European Molecular Biology Laboratory (Heidelberg) have assembled a huge library of genetic information. Since these institutions are supported by public funds, all the information stored there is available to all researchers. In contrast, similar data banks of private firms are not freely accessible, though these firms, too, benefit from public funds. The selection of research subjects by commercial firms depends on the estimates of profit expected to ensue from such research, rather than on considerations of the public good. So, for instance, most of the efforts in genetic engineering are invested in profitable products such as interferon, insulin and growth hormones, which are desirable in developed countries, rather than in developing vaccines or products that are life-saving in developing countries (malaria, hepatitis etc.).[327]

Because of the increasing commercialization of science, especially in the field of biotechnology, genetic engineering and medical research on the one hand, and all the microelectronics-linked industries on the other, more and more scientists have become an integral part of industrial enterprises which thrive on patents, licences, common ventures, takeovers and international sales. The norms of openness, communalism and disinterestedness no longer apply. The advances, inventions, ideas have to be kept secret from others, especially from competitors; the nineteenth century camaraderie and trust of scientists in each other is being replaced by suspicion and often the promulgation of intentionally misleading information.

Nevertheless, I believe that the ideal of open exchange of information should be tenaciously upheld by scientists.

13

How Safe Are Our Drugs?

DRUG TESTING AND FALSIFICATION OF DATA*

Procedures

In the United States Food and Drug Administration (FDA) section of the Department of Health and Human Services (DHHS) is the National Center for Drugs and Biologics (NCDB). One of its divisions in the Office of Drugs, the Division of Scientific Investigations (DSI) (Figure 9) monitors the clinical use of investigational drugs, that is, new drugs that have to be tested before they are approved for widespread use. It is thus the effective tool of the FDA for detecting frauds by the investigators, as well as for exposing misleading statements concerning drugs by their manufacturers or sponsors.

During the period 1964–82 FDA inquiries resulted in some 45 clinical investigators being declared ineligible to receive investigational drugs (12 of them in 1981–2), and an additional six agreed to some restrictions on their investigational work. Some of the disqualified investigators were later criminally prosecuted and sentenced to fines, probation and imprisonment for fraud, fabrication of results, felony etc. Some of these cases will be described in this chapter.

It is important first to understand how new drugs are investigated and approved for use in humans. It is of interest to the manufacturing firms who provide drugs to the investigators to obtain results

---

* Most of the material in this chapter comes from American sources. The Freedom of Information Act enabled me to inspect all the relevant documents pertaining to situations and cases described here.

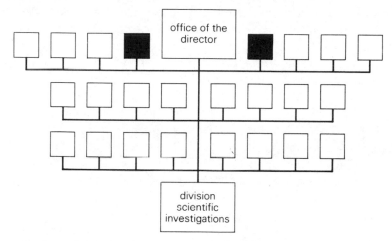

Figure 9. Organizational chart of FDA. Only the two departments relevant to the drug investigations are shown by name.

that would support their request to have the drugs authorized and approved by the FDA; and it is of interest to the clinical investigators to obtain such favourable results, because their clinical testing is paid for by the sponsor, and therefore good results assure the investigator of continued financial support. A study of two dozen patients for two weeks may net the investigator $6500. Several clinical investigators are known to gross more than one million dollars a year for their testing programmes.

In 1980 there were 12,000 clinical investigators conducting clinical trials in the USA[328] (*Wall Street Journal*, 15 May 1980, p. 48). This widespread net leaves some loopholes through which any investigators whose ethical principles are not of the highest quality may slip. Such investigators may omit from their reports results that do not favour the safe clinical use of the drug; or they may improve the clinical figures slightly so as to make them statistically more acceptable to the sponsors. The public must obviously be protected against such unethical practices and the FDA has prescribed a detailed procedure to be followed by the makers and testers of new experimental drugs.

First, the manufacturer has to request an Investigational Drug Exemption (INDE) from the FDA. When this INDE is approved, it allows the manufacturer to send out the drug to his selected investigators (usually physicians). After the investigators have performed the requested studies of the effects of the drug on their patients, they report the results to the sponsor, who in turn seeks from FDA New Drug Approval (NDA) on the basis of the protocols of the tests which are submitted to the FDA. In the NCBD of the FDA a medical officer inspects and examines the data. If he or she (as well as an advisory committee) finds the data satisfactory, the use of a new drug is approved.

The FDA may regulate the actions of the manufacturer or clinical testing of the investigators. FDA may, for instance, find that the explanatory material marketed with the new drug, or the advertisement relating to that drug, is misleading or inadequate. In such cases, regulatory letters may be sent to the relevant sponsor.

Manufacturers

In April 1982, G. J. Gerstenberg, New York District Director of the FDA, informed Nature's County Company in Bohemia (NY) that the antacid digestant tablets distributed by the firm failed to conform to the requirement of over the counter (OTC) antacid drugs since their description 'for rapid relief of heartburn and minor digestive disturbances due to gastric hyperacidity' was a 'false statement in the sense there was lack of substantial evidence of effectiveness and safety derived from adequate and well controlled studies establishing that the formulation of this product is in fact safe and effective for its intended use', and that 'its labeling fails to bear adequate warning' for the protection of the users.

In a similar vein, many firms were warned about the use of a combination drug containing caffeine, phenylpropanolamine and ephedrine, since, though each of the individual drugs was safe, the combination had not been effectively tested.

Another instance of possible misleading information about the nature of a drug's activity concerned an anti-inflammatory and and anti-arthritic drug, benoxaprofen (Oraflex). The producers published a press kit for the information of the public and of the medical profession. In a letter of 27 July 1982, J. A. Halperin, acting Director of the NCDB alerted the company to the possibility that the information accompanying the explanatory tables and figures might be false or misleading to the general public due to 'selective emphasis, inappropriate use of headlines and minimization of adverse information about the drug'. The information in the press kit 'went far beyond the approved labeling for the product into the real or theoretical benefits . . . of unproven clinical significance'. Specifically, the NCDB criticized in some eight points the insinuation by photographs of cells (monocytes) that Oraflex affected them, that side effects were described as 'mild' or 'transient' and that the statement (in the press kit) 'laboratory studies show that [Oraflex] is a potent and specific inhibitor of the migrating inflammatory cells that erode the joints of arthritic patients' was false and misleading, and so were also the headlines such as 'Oraflex–Potential for Suppressing Bone Damage' and 'Medical Research Gains New Insights into Arthritis'. The letter therefore requested the manufacturer to change the wording on his promotional literature accordingly.

Morton Minz, writing in the *Washington Post* on 6 August 1982, stated that although FDA cleared Oraflex for sale in April 1982, 'a check of 173 adverse reactions reports submitted to [name of firm] by five physician testers showed that the company did not tell the FDA of 65 that turned out to be related to the use of the drug'. The policy of the FDA is that selective reporting of adverse reactions is improper and does not serve the public interest.

One of the largest drug and chemical testing outfits in the United States is the Industrial Biotest Laboratories (IBT) in Chicago. It performs more than 12,000 safety tests per year on drugs and chemicals for the industry as well as for the FDA, NCI and the Department of Defense.

In 1976 an FDA pathologist, Adrian Gross, conducted some spot checks on IBT tests performed for the Syntex Corporation on naproxen, a new drug for the treatment of rheumatoid arthritis. Gross found that the number of rats developing tumours and the number that had died were under-reported. He therefore visited IBT and discovered there that some of the dead animals used in the naproxen experiments were not autopsied after death, but disposed of as 'TBD' (too badly decomposed). These findings raised Gross's suspicions, and he requested the FDA in 1977 to arrange an audit of the IBT studies. By the time IBT was subpoenaed by the US Attorney's office in Chicago, many records of the relevant studies had been destroyed by shredding. The president of the company wrote a letter of apology to the US Government for this 'mishap'. The examination of the remaining files by Gross indicated that the data on the study of Triclocarbon (TCC), a deodorant ingredient in soaps, were invalid: new rats were introduced in place of animals that had died, and some rats reported as dying, turned out to be alive later. Tumours found in those animals receiving the chemical were not reported, or euphemistically named 'swollen lymph nodes'. Furthermore, it was found that a former employee of the soap manufacturing firm using the TCC, Paul Wright, worked at IBT on the testing of this compound; his study concluded that TCC did not cause cancer.[329] When the study was completed, Wright was re-employed by his former soap manufacturing firm. In the summer of 1981 this employee and two others were indicted for fraudulent statements in the TCC study. The manufacturing firm provided the legal defence of the accused, and the director of the Environmental Communication Department of the firm minimized the TCC dangers by stating: 'a human would have to eat 26 bars of soap per day for a lifetime to consume the amount of TCC that caused toxic effects in animals'.

In March 1979 the IBT story was brought into the open at Senator Kennedy's Subcommittee on Health and Scientific Research. As a result of these hearings the Naprosyn sales went down by some 20 million dollars. The trial of the four former employees of the soap manufacturer, which was set for March 1982, ended in October 1983 with the IBT officials found guilty of mail fraud* and convicted.[330]

After investigation of some 900 out of 1200 studies on chronic effects of various substances tested by IBT, the examiners on the Environmental Protection Agency (EPA) came to the conclusion that the majority of the reports on these studies were invalid.[328,331] During the hearings of the House Agriculture Subcommittee on Research, chaired by Representative G. Brown, Jnr, the chief pesticide official of the EPA, Edwin Johnson, said that the EPA had notified the manufacturers of 35 compounds that they would have to replace the IBT studies with new data or lose their marketing licence.[332]

Clinical Investigators

A very important regulatory activity of the FDA concerns the clinical investigators of new drugs. When some irregularities are found during the examination of data submitted to the FDA, the Division of Scientific Investigation is alerted. The irregularities discerned may be of several types:

1. Protocol non-adherence: the tests and trials are not performed in exact accordance with the pre-established and approved protocol.

---

* Sending fraudulent data or material by mail or telegraph is a federal offence in the USA.

2. The physician does not maintain accurate case histories and thus some important and relevant information is left out.

3. The investigator fails to submit reports on his tests.

4. The investigating physician fails to obtain informed consent from the patients for the use of the drug.

5. An Institutional Review Board (IRB) organized to follow and review the research on the drug, does not supply a proper approval before the study is begun.

When significant irregularities in these safeguards are detected, the investigator is invited to submit a written statement explaining the reasons for missing or inaccurate information, or he may be invited to informal conferences, or he may be offered a Consent Agreement. If the information provided by the investigator (or sponsor) is still deemed unsatisfactory, a regulatory hearing may be offered. If the Hearing Officer is still not satisfied that the investigation of the drug had been properly done, he may recommend that the FDA declare the investigator ineligible to receive new drugs.

FDA publishes lists of investigators who are either partially or totally restricted in the use of investigational drugs; this restriction is usually limited to a number of years. Following a certain procedure, disqualified investigators may appeal to be reinstated. However, 'studies performed by reinstated investigators prior to the date of disqualification will not be accepted in support of claims before the agency without the agency's acceptance of sponsor validation of such studies' (FDA Release, 15 July 1982).

Some examples of such investigations that led to the disqualification of investigators are presented below.

Dr Harvey Levin, an obstetrician at the Methodist Hospital in Philadelphia and in private practice, submitted false information claiming that he had tested experimental analgesic drugs on 900 maternity patients, when, in fact, he had not given the drug to all of them. These drugs were pain-killers for a surgical procedure designed to widen the birth canal. Levin was disqualified as an investigator for experimental drugs in February 1980 (*Washington Post*, 12 July 1981).

A complicated case of misconduct was brought to light in connection with a study on new drugs sponsored by Bristol Laboratories. The study was conducted at Krest View Nursing Home in Florida in 1974–6 by Dr John H. Close of Miami on some 30 mentally confused patients. The drugs were Librium, phenobarbiton, phenotoin, Triavil, chlorpromazine, flurazepam, etc.

Dr Close had led an impressive career. He had been Medical Research Director of Eaton Laboratories in Norwich (NY), and when he moved to Miami in 1964 he became a fellow of the Mailman Center for Child Development; this academic position enabled him to attract contracts from pharmaceutical firms. In 1974 he established two firms, Med Pharm and Med Slat, and arranged with the Krest Nursing Home that he would conduct the Bristol study. A short time before the FDA investigation of Dr Close's work at Krest Nursing Home (autumn 1977), a charge had been brought against the administration of the home: the previous administrator, Robert D. Wilson, had been accused of not being able to account for the disappearance of some $800,000 of nursing home funds and of faulty administration. When Wilson was replaced by a new administrator, Ms Ruby E. Brown, Dr Close was dismissed for 'tardiness' in the study and for irregularities in his reports to the FDA. Ms Brown was able to provide the FDA investigators with data of only 16 of the nursing home residents allegedly involved in the Bristol study. There was no indication that the investigational drugs were given to any of the patients. Fourteen patients received additional medication which was not justified by their clinical diagnosis and by the experimental protocol; it was not reported to the sponsor and the informed consent of the patients was not obtained. One patient whom Dr Close had included in the study was admitted two months after the study had started, and another was discharged four months prior to the initiation of the study. FDA disqualified Dr Close in July 1978.

Dr Ronald Smith, a psychiatrist, performed studies on psychiatric patients for Sandoz, Lederle and Marion Laboratories during the period 1971–8. Suspicion regarding his work arose when an FDA official on a routine data auditing visit, noted that Dr Smith's consulting room was completely empty except for an executive chair. The official had to sit in a kindergarten chair during the visit![333] Eventually the FDA found that only three out of 60 patients that had supposedly been treated with a mood-changing drug ever received it. According to testimony presented at the Senate hearings, data for the other patients were fabricated by Dr Smith's ex-wife (divorced in 1976): the drugs to be tested were flushed down the toilet. Prior to the investigation, Dr Smith gained wide recognition for his prompt reporting of results at professional meetings; Sandoz even made two training films with Dr Smith portrayed as a model psychiatrist and researcher. He was said to have received $42,000 for his tests on antipsychotic and antidepressant drugs. Dr Smith was disqualified by the FDA, but later provisionally reinstated in 1981.[328]

As a result of regulatory hearings held in the summer of 1979, the FDA disqualified Dr Nathan S. Kline, a prominent New York physician who had been conducting studies on the effect of new neuroactive drugs on psychiatric patients. The drugs were beta-endorphine (14 patients), clozapine, mebanazine (16 patients), lithium carbonate (473 patients) and 1-tryptophan (46 patients). 'Without proper regard to the safety of the patients involved, this study was conducted . . . in an uncontrolled manner and without the benefit of structural design' (Action Memoranda, R. Frankel, Bureau of Drugs, HFD2, 1 August 1979). Counsel for the NCDB sent on 10 August 1979 a letter to the lawyers representing Dr Kline enumerating some 28 instances of violation of FDA regulations. These included the use of beta endorphine (a pain reliever) without an INDE; a false statement to the FDA investigators in May 1978 that Kline had not conducted a study on beta-endorphine, when, in fact, during the preceding year he had presented data on such a study at four scientific meetings; continuation of these studies though notified by the sponsor to cease them; keeping inadequate records; not obtaining informed consent. In the study on mebanazine there were no protocols and no adherence to scheduled laboratory work, and informed consent was not obtained from five out of six patients investigated. With respect to the lithium carbonate study, Kline had no protocols for the study and submitted no reports, and in the case of tryptophan he could not identify patients involved in the study.

Kline did not submit any reports on his studies on clozapine for three years, and when the INDE was terminated he continued to dispense the drug.

During the FDA disqualification hearings of Dr Kline, on 6 September 1979, presided over by Dr Mark Novitch (Associate Commissioner for Health Affairs), the summary stated: 'The inspection has demonstrated a near total lack of concern on Dr Kline's part for our regulations and has disclosed repeated and deliberate violation of these regulations'. At these hearings Dr Kline said: 'Part of my difficulties have arisen simply from literally not following the protocol because the nature of the drug and the reactions dictated looking in some other direction.'

As to his deficiency in not adequately reporting results or not receiving approval for the studies Kline suggested that in studies of this kind a co-investigator might take care of the details of adhering to protocol and of reporting on time. He also commented:

> The ability to recognize where drugs can be used other than the way they are recommended is a very peculiar situation – drugs come, and have done so in the past, from a pharmaceutical company and were based on animal work – they have very specific ideas of what will happen in patients. On at least two or three occasions we have found that the drugs don't do what they [pharmaceutical sponsors] anticipate. But if you have a general enough awareness you recognize that this is the kind of drug that might do something else, for which we are looking.

This argument is valid in animal experimentation, but I doubt whether it can be upheld when it comes to human trials. If novel, unexpected effects come up during an approved trial in patients, the sponsors should be informed and a new protocol should be devised and approved. The arbitrary decision of the investigator to change the type and the course of the treatment cannot be condoned.

Dr Jerome J. Schneyer of Southfield (Michigan) participated in the clinical evaluation of a drug designed for treatment of rheumatoid arthritis and sponsored by Diagnostic Data Inc. When Dr Schneyer submitted to FDA, through his sponsor, serum chemistry data from his patients, the FDA reviewers found them questionable, because they were either the same or very similar in different patients. When in 1977 representatives of FDA asked Schneyer for information about his tests, he refused to grant them access to his laboratory and to patients' records. He thus raised the suspicion that the reported tests had not been carried out. Schneyer claimed that the laboratory tests had been done by a Dr Morita, who died in 1976. The investigators, however, found that the serum chemistry of blood specimens was not done by Morita, because some of the results were dated after Morita's death. Further, an associate of Morita stated that he did not perform tests for Schneyer, who did not even have at his disposal the particular instrument called for in the study protocol. Moreover, the forms recording the results presented by Schneyer were not of the type used by the Beaumont Hospital where Morita worked and the laboratory sheets with the serum chemistry results were not dated or signed by the person supposed to have done the tests. It was later found that the forms were of a type formerly used by another clinical laboratory, but the firm's name and address had been cut off from the bottom of the sheets.

This evidence made it clear to the FDA investigators that Schneyer had supplied them with faked laboratory forms. The inquiry at the Check Up Medical Labs (whose forms Schneyer had used) revealed that no invoice records existed for the relevant periods (the first six months of 1976), either for the physician or for the patients taking the experimental drug. Neither could Schneyer present bills or receipts for money allegedly paid to Dr Morita.

The conclusion reached at this stage was that as Schneyer's results on blood chemistry of patients receiving the experimental drug were not obtained from Morita, nor from the Beaumont Hospital, nor from the Check Up Medical Labs, therefore there were no tests done at all. Schneyer, moreover, refused to provide the investigators with the names and addresses of patients involved in the study.

In the summer of 1977 Schneyer was disqualified by the FDA as a clinical investigator, but the matter did not rest there. He was then prosecuted in court, pleaded guilty on three felony counts and was sentenced to probation and a fine of $12,000.

Another case of prosecution following a disqualification by the FDA concerned Dr Jerome J. Scheiner of Fairfax, Virginia. He reported on the use of six investigational drugs on patients who either did not exist, or were not in the country at the time of the alleged treatment. Scheiner provided the FDA and the sponsors with results of laboratory tests that had not been performed. Scheiner was subsequently indicted on 19 counts arising from extensive falsification of clinical studies and submission of false reports to the government, mail fraud and wire fraud. He pleaded guilty to three counts of fraud and was sentenced to 12 months' imprisonment, $30,000 fines and three years' probation.

The next example presented here involves Dr Fracois Savery of Long Beach, California. Dr Savery worked at the Veterans Hospital as head of the Parkinson's Disease Clinic. In 1975 he conducted a study on Symmetrel for Endo Laboratories and for the Knoll Pharmaceutical Co. (*Wall Street Journal*, 15 May 1980, p. 48). Symmetrel was an antiviral drug which, by chance, had been found to alleviate the symptoms of Parkinson's disease. The FDA investigators who examined his reports noticed that the results of two different studies, submitted to two different sponsors, were identical. There were also inexplicable corrections in handwriting on the laboratory slips. The investigators finally concluded that only two out of six patients reported by Savery had actually received Symmetrel. When asked about the documents relating to the Knoll study, Savery admitted that he had reconstructed them since the originals were allegedly lost when he was in a boat that capsized. Further investigation revealed that Savery had never received a degree of Doctor of Medicine, although he had allegedly obtained it from the University of Saigon. It was also found that he had failed the Massachusetts Medical Boards ten times! After his disqualification by the FDA in June 1978, Savery was brought to court and indicted on eight counts of false documents and two counts of fraud based on fabrication and alteration of the records of some of his laboratory studies. In autumn 1981 Savery pleaded guilty on one count and was put on probation for five years and fined $5000.

The last example concerns Dr Wilbert S. Aronow, who in 1979 was chief of the cardiovascular section of the Veterans Administration Hospital in Long Beach, California, and a member of FDA's advisory committee on cardiology. In the summer of 1982 Aronow resigned from the hospital and joined the University Medical School in Omaha, Nebraska. Simultaneously, he disqualified himself from new drug trials as a result of FDA investigation of his studies on new drugs.

One of these drugs was Minipres (orasozin), an antihypertensive drug claimed by Aronow to be successful also in the treatment of congestive heart failure.[334] A day before the arrival of the FDA investigators, Aronow admitted to the associate director of new drug evaluation that he had changed in his report to the sponsors the radiologist's interpretation of patient's X-rays. The investigators who examined the X-ray film eventually found that only one out of ten X-ray pictures matched Aronow's reports. Aronow signed an affidavit admitting that without even seeing the X-rays he had matched the case reports to what he would have expected if the drug were effective. Aronow also reported that some patients who had frequent attacks of angina greatly improved (one attack per week only) after taking the drug timolol.

What was unusual in this particular case was that the initiative for disqualification came from the clinical investigator himself.

The unwelcome question rears its head: are the documented cases only examples of relatively 'honest' defaulters, not deep enough in misdemeanour to fudge and fabricate the data so completely that they would appear to conform to the FDA requirements? Are there as many wholehearted defaulters eager enough for the support of pharmaceutical firms to have made a much better and undetected job of fabricating their results? The FDA indicates that less than 0.5 per cent of clinical investigators are eventually disqualified for dishonest practices in clinical research. One does not know how large is the number that goes undetected.

THE EXPERTS AND THALIDOMIDE

When newspapers and other mass media try to illuminate a controversial scientific subject, they turn to experts. But who are these experts; how does one become an 'expert' on a scientific subject? One would expect that a scientist who has been working for a number of years in researching the subject in question would be chosen to give his opinion. In reality, however, the choice of the expert is often not related to his having carried out extensive relevant research on the subject or having published the results in a scientific journal. According to June Goodfield[180] the recognition of a scientist as an expert stems from his or her being a head of a recognized agency or institution, or to his or her having been already exposed to the press before.

The scientist who appears as an expert in the public media has grave responsibility. He may be mixing factual data with his own speculations, and this can easily be misunderstood by reporters, who either take the experts' statements at face value, or extrapolate from them.

An example of unfounded expertise is provided by the thalidomide affair. The treatment of pregnant women with this drug in the early 1960s has resulted in 8000 handicapped children in 46 countries. The detailed story of this disaster is told in a book by the Insight Team, of *The Sunday Times*.[335]

Briefly, the developments were as follows. Thalidomide was produced as an analogue of the barbiturates by pharmacist Wilhelm Kunz, working for a then new pharmaceutical company Chemie Gruenenthal. During testing of the new drug reports came in concerning some side effects, such as inflammation of peripheral nerves. The company should have been alerted that thalidomide might be problematic but it denied any causal relationship between the drug and the neuritis, concealing the reports and managing to suppress their publication. These facts were disclosed only later during a court case brought by parents of deformed children against Gruenenthal.

As early as 1960 John Newlinds at the Woman's Hospital in Sydney started trying thalidomide as a general sedative. A leading obstetrician in Sydney, Dr William McBride, also tried this drug; he noticed that three babies born in 1961 suffered the same type of bowel malformation. At the same time Newlinds noticed that the incidence of congenital malformations in infants referred to his hospital was three times the national average, and five times that in a similar women's hospital in Melbourne. Newlinds tried to find the common factor among these cases. He consulted with McBride and found that the only drug all three mothers of the malformed infants he had seen had taken during their pregnancies was thalidomide.

When the first cases of malformation related to thalidomide came to light in England in 1962, the producers of the drug, the Distillers Company (Biochemicals) Ltd, claimed that they should be absolved from legal and moral responsibility because they had performed all the necessary tests known at the time. This statement was borne out by several experts, bringing us back to the consideration with which we began this discussion.

An expert from the Ministry of Health wrote to a complaining parent, Mrs Pat Lane, that 'pharmacologists the world over did not consider it necessary to find out if any new drugs might have such effects' (i.e. produce congenital malformations). A similar statement was issued in the press by Dr Alfred Byrne, a medical correspondent. It later turned out that a statement made by Dr E. Coneybeare, speaking for the Ministry of Health, had been based on information received from his predecessor at the Ministry, who at the time the statement was made worked at Distillers. The medical adviser to Distillers, Dr D. M. Burley, stated that 'no animal test existed at that time to test for the teratogenic properties of the drug'. In fact, all the experts who were consulted at that time were ignorant of the relevant literature.

This literature came to light only in 1971, when a Special Project Unit, headed by Bruce Page, was set up by *The Sunday Times*. A postgraduate medical student, Agnus Maconnachie, was hired to search for relevant literature. He soon found that various pharmaceutical companies in both England and the USA did test their new drugs for possible congenital effects on fetuses and on newborn animals. So, for instance, Wallace Laboratories in the USA, while developing the tranquillizer meprobamate, conducted a comprehensive study on mating male and female rats through gestation, birth and suckling. The paper describing their work was available at the time thalidomide was put on the market.

At Imperial Chemical Industries (ICI) a drug was being developed in 1955 to combat anaemia during pregnancy. ICI had previously (1948) published a paper showing that certain dyes used for the treatment of sleeping sickness produced abnormalities in newborn rats. Other firms who also tested their products for possible reproductive injuries during the decade 1950–60 were: Burroughs-Wellcome (Daraprim), Rhone-Poulenc (chlorpromazine), Hoffman–La Roche (Librium, Valium, Mogadon). Indeed, Hoffman–La Roche had instituted routine reproductive testing of all drugs intended for human use as early as 1944; their example was followed by Pfizer, Lederle and Smith, Kline & French.

Thus, at the time thalidomide appeared on the market the 'experts' should have known that drugs can cross the placental barrier and cause damage to the fetus.

> Though animal tests to prove thalidomide teratogenic were not in general practice before the disaster, tests did exist which would have suggested the dangers to unborn children, because well before thalidomide, it had been established that litter resorption is an indicator of teratogenic potential.[335]

The conclusion reached by *The Sunday Times* team was that from every point of view, scientific as well as moral, Distillers should have done reproductive studies on thalidomide.

The issue was not settled by the 2 years and 7 months long court case brought by parents of deformed children against Distillers. In spite of the case 'a clear legal precedent had not delineated the responsibility of a drug company to its customers'.[180] The matter was finally settled out of court, the company paying damages to the affected children or their families.

THE NCI AND EXPERIMENTAL DRUGS

In autumn 1981 the *Washington Post* (18 October and 4 November) published a series of articles attacking the National Cancer Institute (NCI) for the way it was conducting or sponsoring human trials with experimental anti-cancer drugs. The articles were said to be based on interviews with 600 nurses, doctors, patients and research scientists. Within the scope of these trials over the previous decade some 150 experimental drugs had been given to tens of thousand of cancer patients. The criticism by the press was also based on findings of the FDA investigators that in certain cases unauthorized toxic drugs had been used on terminal cancer patients and that the dosages were too high to be safe. This flurry of accusations was eventually aired before Subcommittee on Investigations and Oversight of the Committee on Science and Technology of the US House of Representatives (Session 1 April 1981).

One of the cases discussed in these hearings involved the use of an anti-cancer drug, methyl-CCNU on cancer-stricken children in 1978. An FDA representative stated at the hearings that the NCI withheld information already available in 1970 in their files and publications indicating that the drug caused kidney damage in dogs and in monkeys; he also stated that since this information was not made sufficiently known to physicians, some 20 children treated for cancer had suffered kidney damage.

Similar criticism was levelled against other experimental drugs such as AMSA, F3TdR, neocarzinostatin, mitoxantrone, piperazinedione, hexamine, 5-aza-cytidine, oestradiol mustard, Adriamycin, nitrosourea, chlorozotocin and maytensine, all of which had known side effects such as nausea, vomiting, mental confusion, and in some cases caused damage to heart and lungs.

The only valid ground for an accusation of unethical behaviour on the part of NCI in these cases is not that these drugs were administered, but that information about the side effects was withheld from the investigating doctors and/or patients. Could such an accusation be substantiated?

At the senate hearing, Vincent de Vita, the director of NCI at that time, put the matter in its proper perspective. He pointed out that each year as many as 40,000 people (that is about 6 per cent of all cancer patients) could be cured by chemotherapy. Therefore, if in a given year there were 1000 drug-related fatalities, this was the price to be paid for the cure of the other tens of thousands of cancer patients. The decision had to be made at what point a drug was promising enough to warrant human testing, and at what point it would be considered toxic and withdrawn; in other words, what was the actual risk/benefit ratio. One has to remember that when the prognosis of cancer is hopeless, and both the doctor and the patient know that suffering, agony and death are the only sure outcome, they are willing to accept higher levels of risk, including the known side effects of the drug. 'The patients are frantically grasping for something and willing to try almost anything in the hope of defeating the cancer, or, at least delaying their death' (Gup and Neumann, *Washington Post*, 18 October 1981, p. A14).

Out of the 150 experimental drugs studied since 1971, de Vita stated, 21 had been thoroughly tested. In 1979 eight of them were recommended by the NCI as 'high priority drugs'. In the meantime, some 40 drugs had accumulated that were known to be useful in the treatment of various forms of cancer. de Vita presented the data for 1980 as follows: out of 785,000 serious cancer patients, 45 per cent are potentially curable – 220,000 by surgery and radiation and 46,000 by chemotherapy with these new drugs.

The criticism voiced by the FDA, in the press and at the Senate hearings led the NCI to form a combined FDA–NCI Investigational New Drug Task Force. This Task Forse was chaired by Lowell T. Harmison. It comprised representatives of the NCI and FDA as well as 13 other persons connected with the Public Health Service. This Task Force studied the matter for three months and in January 1982 published a report which reviewed the prevalent procedures in clinical testing of new drugs, and made specific recommendations for the future. The Task Force reviewed some 70 investigational drugs tested in 68 research centres and universities. During the period 1975–80 eight of these drugs were tested against a number of tumours such as lymphoma, lung, breast and colon carcinomas and on melanomas and leukaemia. The report stated that: 'A number of cases came to the Task Force's attention in which it was alleged that researchers and/or administrators acted improperly in carrying out their responsibilities under the NCI drug development program'.

The Task Force scrutinized two such cases, concerning methyl-CCNU and 5-methyl tetrahydrohomofolate (THHF). Methyl-CCNU was developed in the late 1960s by NCI researchers. Preliminary studies indicated that this drug, though possessing good anti-cancer properties, was also causing acute failure of kidneys in monkeys and in dogs. This information was given in a brochure supplied to the investigators. By 1972 the toxicity (phase I) studies were completed in humans, but no kidney-toxic effects were observed. Similarly, in therapeutic (phase II) studies during the following four years no kidney damage was reported. On the basis of these studies the NCI made Me-CCNU available for wider distribution.

The first report of renal damage, in a five-year-old child receiving the drug for *17 months*, came to the NCI's attention only in August 1977, and this was followed by additional reports in 1978. The NCI contacted all its investigators and detected a few more cases of kidney toxicity.[336] NCI's Adverse Drug Reaction Committee concluded in February 1979 that indeed Me-CCNU was involved in renal insufficiency, and therefore NCI advised all investigators (2020) to consider these findings and to report any further cases. Thus the accusations expressed in the news media and at the Senate hearings, that the information about the kidney damaging properties of Me-CCNU, though available in 1970, had been withheld by the NCI, was discredited by the Task Force. In the mean time the damage was done: the public image of the NCI had been tarnished.

With regard to THHF, the drug was supplied in 1978 to a doctor at the M. D. Anderson Hospital for experimentation *on animals*. In 1980 the investigator and his colleagues carried out pharmacological studies with the drug on six patients, though the Hospital Review Board did not approve such a study. When, in spring 1981, the scientist in charge presented his results on the drug at a meeting, NCI and FDA were alerted and sent a special team for a site visit. They found that the drug had indeed been used on patients. The team recommended the termination of the contract, recovery of the federal funds already spent on these studies, modification of hospital procedures and removal of the investigator involved from leadership in any federally supported research. These recommendations were indeed implemented.

The study of these and other cases led the Task Force to develop recommendations to ensure adequate reporting and monitoring procedures for proper protection of human subjects and for control of investigational drugs. Failure to comply with the recommended procedures would be followed by sanctions.

14

The Pursuit of Honesty

THE EMOTIONAL ROOTS OF CHEATING

Why do people choose a scientific career? In many cases the choice of a new profession is dictated by the desire to follow the footsteps of a parent or relative; in other instances it may be the lure of wealth, influence, or power. Chance also plays an important role in the choice of a career.

Psychological and psychiatric background in the selection of a profession has been discussed by Kubie[337] a psychiatrist with a wide experience of scientists. He proposes that there are neurotic forces that affect the choice and pursuit of a scientific research career. His main argument is that 'scientists as human beings pay a high price for the fact that during the preparation for a life of scientific research their emotional problems are generally overlooked': scientific success cannot solve subconscious personal problems.

When a youth decides upon a choice of career, according to Kubie, he has usually no foreknowledge of what it will be like in reality, because he will have been exposed only to a glorified image of the profession. Even if he has had as an example the career of one of his parents, he still relies on fantasies. Conscious and subconscious forces mingle in the setting of goals to pursue a scientific career (for instance, a subconscious guilt-ridden curiosity about the human body may lead a youth to choose gynaecology or surgery as a profession). Some childhood concerns may assume scientific interest as an acceptable cover.

'Unresolved neurotic anxieties may impel one overanxious young investigator to choose a problem that will take a life time, or alternatively may drive another into easy get-rich-quick tasks which yield an early acclaim . . . '. (Scientists themselves may point to pressures of granting agencies as wielding at least equal influence with regard to choice of problems to be investigated.)

Kubie relates a story of a scientist (known to him as a patient) who by his serious and solid work had proved his case, but his anxieties led him to the falsification of some additional statistical data so as to bolster his already proved hypothesis. Kubie compares such a scientists with a wealthy kleptomaniac: he has everything he needs and wants, but nevertheless is compelled to an action that may land him in jail.

Other examples of Kubie's of the warping of scientific work by the residues of childhood conflicts comprise a brilliant pharmacologist who suffered a latent delusion which affected his choice of hypotheses about drugs and drug action as well as the choice of techniques to prove his theories. Another scientist Kubie describes could never believe his own work; he stole fragments of research from others, even if they supplied unnecessary data, or inserted in his work data which he had made up.

All scientists know that for each successful experiment, one has to do many more that either fail or give inconclusive results. Although the negative experiments are important for the advancement of science, they are not the material on which a scientist can build his (her) career. It is often a matter of chance whether one remains unknown because of negative results, or wins acclaim for positive results. Rewards for industry, perseverance, imagination and intelligence are often accidental: success or failure are determined by chance.

There are some historical examples for the importance of chance: although the vaccination against smallpox was simultaneously discovered by five independent workers, only Jenner gained recognition and honours. The same is true of at least three independent introductions of ether as an anaesthetic, by Crawford W. Long, William T. G. Morton and James Y. Simpson, and eight discoveries of the cellular basis of plant and animal life (Anthony van Leuuwenhoek, Robert Hooke, Nehemiah Grew, Marcello Malpighii, Mathias J. Schleiden, Theodore Schwann, Max J. S. Schultze and Hugo von Mohl).

Our educational system does not warn students, bent on a scientific career, that the creative capacity and willingness to work hard, and even the readiness to make sacrifices to a scientific career by giving up normal family life, recreation and the 'good life', are not *per se* sufficient for achievement of success and recognition. The professional and intellectual training the scientist-to-be receives does not ensure that he or she becomes emotionally prepared for a life of research. Because of the long education period candidates for a research career have to undergo, they have few opportunities to confront themselves with external reality. The educational process may thus select people who subconsciously desire to escape the external reality. For many years the young scientist is economically dependent on fellowships or family support, and this dependence also delays his emotional maturation, characterized by inner assurance, humour and equanimity. The result of the persistence of these unresolved emotional problems may be that when a hard working and even brilliant scientist encounters a series of unsuccessful experiments, which delay his planned and expected progress, he may be tempted to cut corners. Once an experiment is 'helped' to come out right, it is very difficult afterwards to back out of the 'success'. Honesty depends not only on training and education, but ultimately on maturity and security.

In science creativity demands flexibility: a creative scientist should have a high degree of emotional and psychological freedom coupled with an equal degree of organized precision.[338] In the educational system of the Western world, however, everything neurotic in human nature is reinforced. In research the major sources of distortion may be external or internal. The external sources, such as pursuit of money, status or fame, or alleviation of anxiety, may affect, for instance, a physician eager to find a cure, or a pharmaceutical scientist bent on finding a revolutionary drug. The internal sources of distortion are encountered in pure research and, according to Kubie, stem from buried and unresolved neurotic problems.

From my own experience as a university teacher, I know that the attitude of teachers and instructors in exact sciences and in medicine is to impart technical knowledge to the students; to teach them problem-solving stratagems and methods for acquiring professional information. Any emotional or psychological problems the students might have are dealt with on an ad hoc basis when they arise. There is no systematic attempt to evaluate individuals wishing to embark on a scientific career as to their mental and emotional make-up.

I am not sure whether a character trait of absolute honesty is a prerequisite or the driving force for choosing a scientific career.

SAFEGUARD MECHANISMS IN SCIENCE

As our study of scientific frauds and forgeries draws to an end, it is time to try to arrive at some conclusion as to how serious is the occurrence of fraud in scientific research. In the opinion of Dr Ronald Lamont-Havers, Director of Research of the Massachusetts General Hospital, while testifying before the Congressional Subcommittee on Investigations and Oversight (31 March 1981) in the case of Dr Long (see chapter 6), the incidence of deliberate falsification of scientific data is rare, and the apparent increase in the number of incidents in the past decade is due to increased reporting of these incidents by the lay as well as by the scientific press.

One of the most common types of fraud discussed at the hearings of this Subcommittee was the misuse of funds allocated to the researcher by a public or governmental granting agency. This is mostly due to discrepancy in the interpretation of the rigid regulations governing the use of the funds.

The other type of fraud is the actual falsification of data and rigging of experiments. Those who commit such misdeeds must surely be acting on the basis of irrational impulses, since any rational scientist knows that an intentional falsification of data to fit a hypothesis will sooner or later be found out by others. A misjudgement in this respect on the part of an unethical scientist seems to be compounded by the situation that most research nowadays is done in collaborative groups headed by a principal investigator (Pl) (usually a senior scientist). Because of the splitting of specialties the Pl is obliged to rely on the integrity and honesty of his collaborators in those fields in which he himself is not an expert.

It often happens that scientists rush into publication with a new finding by reporting it in the form of an abstract, including in it hopeful rather than actual results, assuming that by the time they will have to present their data at a conference or a meeting, the data will be already available. Lamont-Havers has a mitigating view about errors resulting from such conduct; he calls this 'misdirected enthusiasm'.

As to the peer review system, which is supposed to be the watchdog of propriety in publication, one must remember that the committees and study sections are also composed of mere people and thus constitute a 'very human system with all of the idiosyncrasies, biases, and imperfections which make us human'. On the other hand, as long as the composition of these committees, boards or study sections is dictated by scientific merit only, and not by some other irrelevant criteria such as sex, colour, race or institutional affiliation, the decisions arrived at will be fairly just and competent. 'The peer self-examination review and criticism of a scientific system which allows freedom of inquiry is by far the most effective means by which error can be detected (Lamont-Havers). The peer review and the editorial censorship of papers submitted for publication ensure that, in spite of some mishaps, the whole enterprise is kept at a very high level of competence. B. Davis, in his article on 'Neo-Lysenkosim, IQ and the press' wrote:

> The strength of science in analysing reality comes from its strict separation of facts from values, of observations from expectations . . . the key to success in scientific enterprise is its passionate dedication to objectivity. Its advance depends on accepting the conclusions dictated by verifiable observations and by logic, even when they conflict with common sense or with treasured preconceptions. [339]

Broad and Wade[2] claim that safeguard mechanisms in science, such as peer review of grant applications, refereeing of papers submitted for publication and attempts to replicate experiments in other laboratories, do not work adequately in curbing the extent of misconduct in science. They argue that cheating is practically never detected by peer reviewers or referees, and only a few scientists bother properly to replicate experiments of others.

This claim is only partially true. Norton Zinder, Professor of Microbial Genetics at the Rockefeller University, countered these claims at a meeting of the Association for the Advancement of Science held in Detroit in 1983. Science would not have survived if cheating were as prevalent as Broad and Wade claim in their book, Zinder declared.[340] The fact that there are fraudulent scientists explains nothing about the structure of science; it only teaches us something about an individual, marginal scientist: 'From fraud one learns only about fraud'.[340] Some of the most notorious cases of fraud that gained wide publicity involved 'mentally exhausted scientists', like Summerlin, or 'psychopaths' (Spector, Darsee) who would have cheated in any profession they entered. Zinder believes that the checking mechanisms (peer review and refereeing) fulfil their intended function, though they may fail on occasion. Introduction of a more efficient mechanism, says another author, Ziman, would destroy one of the mainsprings of scientific community, namely the trust in each other's integrity: 'A fierce and uncompromising honesty is one of the standard attributes of the so-called "scientific attitude" . . . it is certainly moulded into one by powerful social pressures in the graduate student phase of one's career.'[341]

The views of Broad and Wade have been reviewed and challenged in a number of periodicals. For example, Alan R. Price, Associate Professor of Biological Chemistry at the University of Michigan[342] considered some of their statements as too sweeping if not actually untrue. Consider this passage from their book:

> Cases of fraud provide telling evidence not just about how well the checking systems of science operate in practice, but also about the fundamental nature of science – about the scientific methods, about the motives and attitudes of scientists . . . the roots of fraud lie in the barrel, not in the bad apples that occasionally roll into public view.[2]

Fraudulent behaviour by clever and intelligent scientists, designed to circumvent the normal checks and balances operating in science, cannot be taken as a base upon which to judge 'the fundamental nature of science'; a hold-up in a bank or cracking of a safe does not explain in any way the nature and the principles of the banking system. All it does is permit us to evaluate whether there exist sufficient methods to prevent or abort any further attempts at similar misconduct.

Though the claim of Broad and Wade that there is a correlation between fraud and the quality of the master–apprentice relationship may well be true, their estimate of its extent seems to be exaggerated. The number of cases where the special trust that mentors put in their graduate or postgraduate students and in their colleagues has led to the abuse of this trust is very small in proportion to the tens and perhaps hundreds of thousands of relationships involving senior and junior scientists that exist in modern science.

It is not only the official world of published accomplishments on which scientific relationships and communication are based. There is also a much wider network of informal communication in science, which further blocks attempts at promulgating falsehoods, and which in most cases casts out the inept, dishonest individuals before they become fully fledged scientists.

In practically every conversation I have had on this subject with heads of laboratories, professors and teachers, they could report attempted cases of falsification or cheating which they had exposed early enough to get rid of the offending individual and bar his access to the halls of science. These incidents are known about but are not usually reported, unless over a cup of coffee or at a bar. They simply provide a memento that dishonest people can be found in every walk of life.

Alfred Meyer, co-author with Norton Zinder, says:

> Contrary to the main premise of the book [Broad and Wade] there are also individuals of sterling integrity, in particular those who honored the doubts within themselves and proceeded, very often at risk, to question the results of colleagues and superiors – in short as many heroes as goats.[340]

One should remember, after all, that the world produces few saints.

The remedy which Wade advocates, to which I also subscribe, is the creation of an atmosphere in science that would reduce the temptation to fraud, that would take care of a more honest distribution of credits, reduce the number of vanity press journals and raise research standards.[343]

15

What Can Be Done About Fraud in Science?

If one has detected a fraud in science, knowing how best to tackle it can be a problem. We have witnessed in the Lucas case (chapter 7) that the course of events which eventually led to Lucas's downfall began with a complaint first lodged to the university authorities and then to a law court by a 'post-doc' employee who was fired. Darsee (chapter 6) was denounced by his technicians. Many other cases of misconduct are detected by scientific or technical employees, students and junior colleagues who witness unethical behaviour on the part of their superiors or colleagues.

The proper action to take by a lower ranking scientist or technician would be to approach the offender, point out the problem and urge correction. In many administrative situations this means sticking one's neck out and facing the possibility of losing a job. If the superiors fail to respond, the employee may bring the matter before the proper authorities or even before the public. Such action demands civic courage, and people embarking on it are often 'whistle-blowers'.[344]

Whistle-blowing may, however, land those who practise it in deep trouble. A few years ago Clifford Richter, a health physicist at a hospital in Columbia, Missouri, reported some violations of safety regulations to the Nuclear Regulatory Commission, as was his duty. The hospital abolished the man's job. One would think that people exposing fraud, or warning the public about wrong-doings they had encountered, would be heard, fairly judged and would deserve commendation or even promotion; certainly not discharge. The possibility exists, however, that the whistle-blowers may be wrong:

they may have a personal grudge, or be out to get vengeance, or they may be cranks, or they may simply be mistaken.

Constructive dissent has not yet received its definition and procedure for due process, for institution of formal hearings, for the possibility of an appeal. The drive for getting at the truth should be the dominant passion of a responsible scientist, but there is something lacking in the definition of 'truth'.

Morris B. Baslow was a marine biologist employed since 1974 as a senior scientist by Lawler, Matusky and Skelly Engineers. His task was to conduct research on the effect of thermal effluents from a power plant on fish in the Hudson river. The Environmental Protection Agency (EPA) had asked the plant (Consolidate Edison) to instal cooling towers to reduce the amount of hot water being released into the river. Edison contracted Lawler, Matusky and Skelly Engineers to study the matter and to demonstrate that thermal effluents from the plant were not damaging the river's aquatic life. Baslow, however, found that larval and fish growth were dependent on *optimal* temperature, and that any temperature above the desired range, inhibited growth. He tried to convince his employers to include this data in their testimony at the EPA hearings. Since his superiors did not agree with his point of view, Baslow sent a letter to the administrative law judge Thomas B. Yost, accusing his company of 'perjury' and of presenting invalid data about the 'density dependent growth' of aquatic life. This action resulted in Baslow's being dismissed from his work in 1979. He filed a suit with the Department of Labor claiming that under the Federal Water Pollution Control Act he should have been protected against dismissal following his revelations.

The company responded with a suit against Baslow demanding the return of documents Baslow had removed from the laboratory, alleging that Baslow's actions were defamatory and asking for some five million dollars in damages. Late in 1980 the case was settled out of court: Baslow agreed to return the documents and to apologize for the use of the word 'perjury'.

Baslow's case, as described by Constance Holden,[345] 'has amply demonstrated what a nasty – and time consuming – imbroglio a little simple whistle-blowing can create'.

The matter of whistle-blowing was reviewed in 1981 by the State of Michigan. A Whistle Blowers Protection Act was passed, which applied to any private and public employees. Any employee who reports what in his belief is a violation of law by the employer, and as a result of this action is then dismissed or punished, can now sue the employer in a public court. If the employer cannot prove that the dismissal or the punishment were for valid business or personal reasons, the court can award the employee damages and legal costs, order the reinstatement of the employee, and in addition impose a fine on the employer.

Such a course of action may be proper in business of industry, but it is not as simple in the research and academic world. What can the university administration do in a case of whistle-blowing so as to be fair and just to both the accuser and the accused?

The following steps are advocated by Professor Alan F. Westin of Columbia University:[346] The university can draw up a code of legal rules and norms for the pursuit of research activities, and commit its administration to observing and executing these rules. He recommends that an institutional review board be established, whose function would be the examination of grant proposals, submitted to the government, as well as identification of ethical issues involved. In addition to the review board, there should be an appeals committee at the university, to which complaints about research misconduct could be made. This committee should also insure that there would be no reprisals against any person who filed a complaint.

Further discussion of the issues raised by whistle-blowing may be found in a paper by Edsall.[347]

PREVENTION

In 1982 the executive council of the Association of American Medical Colleges (AAMC) published a guide to ethical standards in science for medical schools and teaching hospitals.[348, 349] The prime responsibility for prevention of fraud in research is laid on the scientific community itself and on the academic institutions. Though some cases of fraud cannot be entirely prevented, it is held that faculties can create a climate that promotes high ethical standards.

Each institution has to define clearly the rules and procedures to be implemented when suspicion of fraud is raised. These rules must also encompass the provision of adequate opportunities for suspects to defend themselves. One of the rules which the AAMC committee insisted on was that all the suspects' research activities should be reviewed, not only the incident on which the allegation of fraud was based. Once an alleged fraud is substantiated, the sponsoring agency should be notified so it can take appropriate action (e.g. restitution of funds). All pending papers and abstracts related to the fraudulent research should be withdrawn, and the editors of the journals concerned notified. The institution employing the fraudulent researcher should terminate his or her employment and take action against any other faculty members whose misconduct was substantiated. If the alleged fraud is not substantiated, steps should be taken to restore the reputation of the researcher and others involved in the investigation. Action should also be taken against the parties who levelled the unjust or dishonest accusations.

The first university to announce a plan for dealing with fraudulent research or deceit was Yale. Its new plan addresses the situation where research results are published by a number of collaborators, one of whom has supplied fraudulent data. The university unequivocally states: 'By claiming authorship of a scholarly publication, each collaborator must accept the discredit, as well as the credit, for collaborative effort'.[350]

In June 1982 the NIH held a staff symposium to develop a unified policy on misconduct in research. It was recognized that fraud is committed if a false statement or fake data are submitted either to the granting agency or to a journal. Either case should be considered to be within the jurisdiction of an appropriate federal government department, because 'the sharing of accurate scientific data generated by federal biomedical research dollars among scientists is a matter in which the government has the power to act'.[351]

It seems certain that other research institutions and universities will adopt similar plans for controlling unethical behaviour. The ultimate safeguards, however, must reside in the vigilance of the scientific community. It must ensure that any fraud or dishonesty in scientific research, particularly in fields important to the advancement of science, is discovered by its irreproducibility within a brief time. Offenders must be left in no doubt that transgression will promptly bring their careers to an ignominious end.

To understand deceit in science one has to study first the ethics of the society within which the scientists work and live, ethics which Professor Michael Riese of the University of Guelph (Ontario) has defined as a 'shared illusion of the human race'.

' . . . if you have not asked yourself whether your work as a scientist is committing you to activities which you would not engage in as a lay person, then you are not only avoiding your civic responsibilities – you are behaving in a manner which is incompatible with being a good scientist'.[352]

References

1 Zuckerman, H. 1977: Deviant behavior and social control in science. In E. Sagarin (ed.) *Deviance and Social Change*, Beverly Hills: Sage Publications.

2 Broad, W. and Wade, N. 1982: *Betrayers of the Truth. Fraud and Deceit in the Halls of Science*. New York: Simon and Schuster.

3 Merton, R. 1973: The normative structure of science. In N. Storer (ed.) *The Sociology of Science. Theoretical and Empirical Investigations*, Chicago: Chicago University Press.

4 Barber, B. 1952: *Science and the Social Order*. New York: Free Press.

5 Cournand, A. and Zuckerman, H. 1975: The code of science. In P. Weiss (ed.), *Knowledge in Search of Understanding*, Mt Kisco, New York: Publishing, ·126–47.

6 Mohr, M. 1979: The ethics of science. *International Science Reviews*, 4: 45.

7 Gaston, J. 1978: Disputes and deviant views about the ethos of science. In *The Reward System in British and American Science*. New York: Wiley, 158–84.

8 Dixon, B. 1984: Deniers of the truth. *The Sciences*, 24: 23.

9 Jevons, F. R. 1973: *Science Observed*. London: Allen and Unwin.

10 Kleinbaum, D. C., Morgenstern, H. and Kupper, L. L. 1981: Selection bias in epidemiological studies. *American Journal of Epidemiology*, 113: 452.

11 Weber, J. 1969: Evidence for the discovery of gravitational radiation. *Physical Review Letters*, 22: 1320.

12 Saint John-Roberts, I. 1976: Are researchers trustworthy? *New Scientist*, 71, 2 September: 481.

13 Saint John-Roberts, I. 1976: Cheating in science. *New Scientist*, 72, 26 November: 465.

14 Roberts, R. R. 1977: An unscientific phenomenon. Fraud grows in laboratories. *Science Digest*, June, p. 38.

15 Wolins, L. 1962: Responsibility for raw data. *American Psychologist*, 17: 657.
16 Zahler, R. 1976: Errors in mathematical proofs. *Science*, 193: 98.
17 Luria, S. A. 1975: What makes a scientist cheat? *Prism*, May, p. 16, 18.
18 Weiss, P. 1964: Science and the university. *Daedalus*, 93: 1184, 1211.
19 Rosenthal, R. 1966: *Experimental Effects in Behavioral Research*. New York: Appleton–Century–Croft.
20 Hyman, H. H., Cobb, W. J., Feldman, J. J., Hart, C. W. and Stember, C. H. 1954: *Interviewing in Social Research*. Chicago: University of Chicago Press, 4.
21 Rosenthal, R. and Fode, K. L. 1963: The effect of experimenter bias on the performance of the albino rat. *Behavioral Science*, 8: 183.
22 Rosenthal, R. and Lawson, R. 1964: A longitudinal study on the effects of experimenter bias on the operant learning in laboratory rats. *Journal of Psychiatric Research*, 2: 61.
23 Bernstein, L. 1957: The effect of variations in handling upon learning and retention. *Journal of Comparative Physiology and Psychology*, 50: 162.
24 Hetherington, N. S. 1983: Just how objective is science? *Nature*, 306: 727.
25 van Maanen, A. 1916: Internal motion in the spiral nebula, Messier. *Astrophysical Journal*, 44: 210.
26 van Maanen, A. 1923: Investigation on proper motion: Internal motion of the spiral nebula, Messier 33, NGC 598. *Astrophysical Journal*, 57: 264.
27 Hetherington, N. S. 1975: The simultaneous 'discovery' of internal motions in spiral nebulae. *Journal of History of Astronomy*, 6: 115.
28 Hubble, E. 1935: Angular rotations of spiral nebulae. *Astrophyscial Journal*, 81: 334.
29 Stenflo, J. 1970: Hale's attempt to determine the sun's general magnetic field. *Solar Physics*, 14: 263.
30 Rostand, J. 1960: *Error and Deception in Science*. New York: Basic Books, 13–29.
31 Langmuir, I. (edited by R. N. Hall) 1985: Pathological science (scientific studies based on non-existent phenomena). *Speculations in Science and Technology*, 8: 77.
32 Barnes, A. H. and Davis, B. 1930: The capture of electrons by alpha particles *Physical Review*, 35: 215.
33 Webster, H. C. 1930: Capture of electrons by alpha particles. *Nature*, 126: 352.
34 Davis, B. and Barnes, A. H. 1931: Capture of electrons by swiftly moving alpha particles. *Physical Review*, 37: 1368.
35 Allison, F. 1927: The effect of wavelenfth or the difference in the lags on the Faraday effect behind the magnetic field for various liquids. *Physical Review*, 30: 66.

36 Allison, F. and Murphy, F. J. 1930: A magnetic optic method of chemical analysis. *Journal of the American Chemical Society*, 52: 3796.

37 Allison, F. and Murphy, F. J. 1930: The probable number of isotopes of eight metals determined by a new method. *Physical Review*, 36: 1097.

38 Bishop, E. R., Lawrenz, M. and Dollins, C. B. 1933: Lead isotopes. *Physical Review*, 33: 43.

39 Allison, F. and Bishop, E. P. 1933: Bismuth isotopes. *Physical Review*, 43: 47.

40 Latimer, W. H. and Young, H. A. 1933: Isotopes of hydrogen by the magneto-optic method. The existence of H^3. *Physical Review*, 44: 690.

41 McGhee, J. L. and Lawrenz, M. 1932: Tests for the element 87 (Virginium) by the use of Allison's magneto-optic apparatus. *Journal of the American Chemical Society*, 54: 405.

42 Jeffesen, M. A. and Bell, R. M. 1935: An objective study of Allison Magnetic-optic method of analysis. *Physical Review*, 47: 546.

43 Gurwitch, A. 1923: Die Natur des spezifischen Erregers der Zellteilung. *Archiven fuer gesammte Entwicklungsmechanik*, 100: 11.

44 Gurwitch, A. 1932: *Die Mitogenetische Strahlung*. Berlin: Springer-Verlag.

45 Wolff, L. K. and Ras, G. 1932: *Biochemische Zeitschrift*, 250: 305.

46 Hollaender, A. and Claus, W. D. 1935: Some phases of the mitogenic ray phenomenon. *Journal of the Optical Society of America*, 23:270.

47 Taylor, G. W. and Harvey, E. N. 1931: The theory of mitogenetic radiation. *Biological Bulletin*, 61: 280.

48 Lorenz, E. 1934: Search for mitogenetic radiation by means of the photoelectric method. *Journal of General Physiology*, 17: 843.

49 Rylska, T. 1948: The mitogenetic radiation of growing yeast and of potato tubers. *Annals of the University Marie Curie Sklodowska*, section C, Biology 3, 335.

50 Moiseva, M. N. 1960: Promitohenetychneprominya [Mitogenetic radiation] *Ukrainske Botanichniy Zhurnal*, 17: 29.

51 Gurwitch, A. A. and Eremeev, V. F. 1966: Mitogenetic radiation of the myocardium of animals (Russian). *Bulletin of Experimental Biology and Medicine*, 61: 56.

52 Gurwitch, A. A., Eremeev, V. F. and Sobieva, Z. I. 1966: (Russian) *Bulletin of Experimental Biology and Medicine*, 62: 55.

53 Hoefert, M. 1968: Schwaches Leuchten aus Organismen und mitogenetische Strahlung. *Strahlentherapie*, 135: 103.

54 Fedyakin, N. N. 1962: Change in structure of water during condensation in capillaries. *Colloid Journal* (USSR), (English trans.) 24: 425.

55 Deryaguin, B. V. and Churayev, N. V. 1971: Investigations of the properties of water II. *Journal of Colloid Interface Science*, 36: 415.

56 Lippincott, E. R., Stromberg, R. R., Grant, W. H. and Cessac, G. 1968: Polywater, vibrational spectra indicate unique stable polymeric structure. *Science*, 164: 1482.

57 Linnet, J. W. 1970: Structure of polywater. *Science*, 167: 1719.

58 Allen, L. C. 1971: An annotated bibliography for anomalous water. *Experientia*, 36: 554.

59 Hildebrand, J. H. 1970: 'Polywater' is hard to swallow. *Science*, 168: 1397.

60 Rousseau, D. L. and Porto, S. P. S. 1970: Polywater: Polymer or artifact? *Science*, 167: 1715.

61 Rousseau, D. L. 1971: An alternative explanation for polywater. *Journal of Colloid Interface Science*, 36: 434.

62 Stromberg, R. R. and Grant, W. H. 1971: Polywater – a search for alternative explanation. *Journal of Colloid Interface Science*, 36: 443.

63 Kurtin, S. L., Mead, C. A., Mueller, W. A., Kurtin, B. C. and Wolf, E. D. 1970: Polywater – A hydrosol? *Science*, 167: 1720.

64 Allen, L. C. 1971: Theoretical evidence against the existence of polywater. *Nature*, 253: 550.

65 Allen, L. C. 1971: What can theory say about the existence and properties of anomalous water? *Journal of Colloid Interface Science*, 36: 469.

66 Allen, L. C. 1973: The rise and fall of polywater. *New Scientist*, 59: 376.

67 Franks, F. 1980: *Polywater*, Cambridge, Mass.: MIT Press.

68 Byrne, W. L. *et al.* (16 authors) 1966: Memory transfer. *Science*, 153: 658.

69 Ungar, G., Galvan, L. and Clark, R. H. 1968: Chemical transfer of learned fear. *Nature*, 217: 1259.

70 Ungar, G., Desiderio, D. M. and Parr, W. 1972: Isolation, identification and synthesis of a specific behavior inducing brain peptide. *Nature*, 238: 198.

71 Stewart, W. W. 1972: Comment on the chemistry of scotophobin. *Nature*, 238: 202.

72 Ungar, G. 1970: *Molecular Mechanisms in Memory and Learning*. New York: Plenum Press.

73 Parr, W. and Holzer, G. 1972: Zur Struktur von Skotophobin. *Experientia*, 28: 884.

74 Ali, A., Faesel, J. H. R., Sarantakis, D., Stevenson, D. and Weinstein, B. 1971: Synthesis of a structure proposed for scotophobin. *Journal of Colloid and Interface Science*, 27: 1138.

75 de Wied, D., Sarantakis, D. and Weinstein, B. 1973: Behavioral evaluation of peptide related to scotophobin. *Neuropharmacology*, 12: 1109.

76 Miller, R. R., Small, D. and Berk, A. M. 1975. Information content of rat scotophobin. *Behavioral Biology*, 15: 463.

77 Wojcik, M. and Niemierko, S. 1978: Effect of synthetic scotophobin on motor-activity in mice. *Acta Neurobiologiae Experimentalis*, 38: 25.

78 Martinez, J. L. (ed.), 1981: *Endogenous Peptides and Learning and Memory Processes*. New York: Academic Press.

79 Neugebauer, O. 1975: *A History of Ancient Mathematical Astronomy*. New York: Springer-Verlag.

80 Gingerich, O. 1976: Review of Neugebauer's book: A history of ancient mathematical astronomy. *Science*, 193: 476.

81 Westfall, R. S. 1973: Newton and the fudge factor. *Science*, 179: 751.

82 Boultree, A.H. 1973: The fudge factor. *Science*, 180: 1121.

83 Manuel, F. E. 1976: *A Portrait of Isaac Newton*. Cambridge, Mass.: Harvard University Press, 178.

84 Bennet, J. H. (ed.) 1965: *Experiments in Plant Hybridization by G. Mendel*. Edinburgh: Oliver and Boyd.

85 Fisher, R. A. S. 1936: Has Mendel been rediscovered? *Annals of Science*, 1: 115.

86 Sturtevant, A. H. 1965: *A History of Genetics*. New York: Harper and Row.

87 Orel, V. 1968: Will the story of 'too good' results of Mendel's data continue? *Bioscience*, 18: 776.

88 Iltis, H. 1932: *Life of Mendel*. New York: W. W. Norton.

89 Dunn, L. C. 1965: *A Short History of Genetics*. New York: McGraw–Hill.

90 Roberts, H. F. 1929: *Plant Hybridization before Mendel*. Princeton: Princeton University Press, 292.

91 Gardner, M. 1977: Great fakes of science. *Esquire*, October, pp. 88–92.

92 Zirkle, C. 1954: Citation of fraudulent data. *Science*, 120: 189.

93 Campbell, M. 1965: Explanation of Mendel's results. *Centaurus*, 20: 159.

94 Pearl, R. 1940: *Introduction to Medical Biometry and Statistics*, 3rd edn Philadelphia: Saunders, 85–9.

95 Root-Bernstein, R. S. 1983: Mendel and methodology, *History of Science*, 21: 275.

96 Stern, C. and Sherwood, E. R. (eds) 1966: *Origins of Genetics. A Mendel's Source Book*. San Francisco: London, 39–40.

97 Gey, G. O. 1933: An improved technic for massive tissue culture. *American Journal of Cancer*, 17: 752.

98 Jungenbluth, C. W. 1937: Vitamin C therapy and prophylaxis in experimental poliomyelitis. *Journal of Experimental Medicine*, 65: 127.

99 Jungenbluth, C. W. 1937: Further observations on vitamin C therapy in experimental poliomyelitis. *Journal of Experimental Medicine*, 66: 459.

100 Sabin, A. 1939: Vitamin C in relation to experimental poliomyelitis. *Journal of Experimental Medicine*, 5: 507.

101 Kihara, H. 1952: T. D. Lysenko. *Science and Culture*, 17: 462.

102 Kammerer, P. 1923: Breeding experiments on the inheritance of acquired characters. *Nature*, 111: 637.

103 Kammerer, P. 1923: Breeding experiments on the inheritance of acquired characters. *Nature*, 112: 1237.

104 Kammerer, P. 1924: *The Inheritance of Acquired Characters*. New York: Boni and Liveright.

105 Bateson, W. 1923: Dr Kammerer's *Alytes*. *Nature*, 111: 738.

106 Bateson, W. 1923: The inheritance of acquired characters in *Alytes*. *Nature*, 112: 391.

107 Bateson, W. and Przibram, H. 1923: Experiments on *Alytes* and *Ciona*. *Nature*, 112: 899.

108 McBride, E. W. 1923: Dr Kammerer's experiments. *Nature*, 111: 841.

109 Noble, G. K. 1926: Kammerer's *Alytes*. *Nature*, 118: 209.

110 Przibram, H. 1926: Kammerer's *Alytes*. *Nature*, 118: 210.

111 Kammerer, P. 1926: Paul Kammerer's letter to the Moscow Academy *Science*, 64:93.

112 Koestler, A. 1972: *The Case of the Midwife Toad*. New York: Random House.

113 Fothergill, P. G. 1953. *Historical Aspects of Organic Evolution*. New York: Philosophical Library, 253–8.

114 Mason, S. F. 1953: *Main Currents in Scientific Thought*. New York: Henry Schuman, 437.

115 Segal, J. 1951: *Michurin, Lysenko et le Probleme de l'Heredité*. Paris: F. Réunis.

116 Gould, S. J. 1972: Zealous advocates. (Review of Koestler's 'The Case of the Midwife Toad'). *Science*, 176: 623.

117 Koestler, A. 1964: *The Act of Creation*. New York: Macmillan.

118 Witkowski, J. A. 1979: Alexis Carrel and the mysticism of tissue culture. *Medical History*, 23: 279.

119 Witkowski, J. A. 1980: Dr. Carrel's immortal cells. *Medical History*, 24: 128.

120 Carrel, A. 1912: On the permanent life of tissue outside of the organisms. *Journal of Experimental Medicine*, 15: 516.

121 Ebeling, A. H. 1922: A ten year old strain of fibroblasts. *Journal of Experimental Medicine*, 35: 755.

122 Fischer, A. 1925: *Tissue Culture*. Copenhagen: Levin and Munkgaard.

123 Hayflick, L. 1956: The limited *in vitro* lifetime of human diploid cell strains. *Experimental Cell Research*, 37: 614.

124 Hayflick, L. 1976: The cellular basis for biological aging. In C. E. Finch and L. Hayflick (eds) *The Biology of Aging*, New York, Van Nostrand, 159–86.

125 Hayflick, L. and Moorehead, P. S. 1961: The serial cultivation of human diploid cell strains. *Exp. Cell Res.* 25: 585.

126 Medawar, P. B. and Medawar, J. S. 1977: The life science. *Current Ideas in Biology*, pp. 125–6.

127 Burt, C. 1955: The evidence for the concept of intelligence. *British Journal of Educational Psychology*, 25: 158.

128 Burt, C. 1958: The inheritance of mental ability, *American Psychologist*, 13: 1.

129 Herrnstein, R. 1973: *I.Q. in the Meritocracy*. Boston: Little, Brown.

130 Burt, C. 1966: The genetic determination of differences in intelligence. A study of monozygotic twins reared together and apart. *British Journal of Psychology*, 57: 137.

131 Herrnstein, R. 1971: I.Q. *Atlantic Monthly* 228: 43.

132 Kamin, L. 1974: *The Science and Politics of IQ*. Potomac: Erlbaum.

133 Burt, C. 1939: The relations of educational abilities. *British Journal of Educational Psychology*, 9: 45.

134 Burt, C. 1943: Ability and income. *British Journal of Educational Psychology*, 13: 93.

135 Friedman, D. M. 1954: Early researches on general and special abilities. *British Journal of Statistical Psychology*, 7: 119.

136 Burt, C. 1954: Experimental tests of higher mental processes and their relation to intelligence. *Journal of Experimental Pedagogy* I, 1911, 93–112.

137 Jensen, A. R. 1974: Kinship relations reported by Sir Cyril Burt. *Behavior Genetics*, 4: 1.

138 Eysenck, H. J. 1972: Obituary, Sir Cyril Burt (1883–1971). *British Journal Mathematical and Statistical Psychology*, 25: iv.

139 Kamin, L. 1980: Burt, Burt he hasn't a shirt. *New Scientist*, 88,20 November: 498.

140 Burt, C. and Howard, M. 1956: The multifactorial theory of inheritance and its application to intelligence. *British Journal of Statistical Psychology*, 9: 95.

141 Burt, C. and Howard, M. 1957: The relative influence of heredity environment on assessment of intelligence. *British Journal of Statistical Psychology*, 10: 99.

142 Conway, J. 1958: The inheritance of intelligence and its social implications. *British Journal of Statistical Psychology*, 11: 171.

143 Wade, N. 1976: IQ and heredity: suspicion of fraud beclouds classic experiment. *Science*, 194: 916.

144 Dorfman, D. D. 1978: The Cyril Burt Question: New Findings. *Science*, 201: 1177.

145 Burt, C. 1961: Intelligence and social mobility. *British Journal of Statistical Psychology*, 14: 3.

146 Spielman, W. and Burt, C. 1926: *A Study of Vocational Guidance*. London: HMSO, 12–17.

147 Erlenmeyer-Kimling, L. and Jarvik, L. F. 1963: Genetics and intelligence. A review. *Science*, 14: 1478.

148 Estling, R. 1982: Estimation of Burt. *New Scientist*, 95, 1 July: 47.

149 Edwards, J. H. 1982: Estimation of Burt. *New Scientist*, 94, 17 June: 803.

150 Fletcher, H. 1982: My work with Millikan on the oil drop experiment. *Physics Today*, June, p. 43.

151 Millikan, R. A. 1910: The isolation of an ion, a precision measurement of its charge, and correction of Stokes' Law. *Science*, 32: 436–48.

152 Holton, G. 1978: Subelectrons, presuppositions and the Millikan-Ehrenhaft dispute. *Historical Studies in Physical Sciences*, 9: 166.

153 Franklin, A. D. 1981: Millikan's published and unpublished data on oil drops. *Historical Studies in Physical Sciences*, 11: 185.

154 Millikan, R. A. 1923: Science and Society, *Science*, 58: 293.

155 Gregory, F. G. and Purvis, O. N. 1938: The phenomenon of vernalization, the acceleration of ear formation in winter varieties of cereals by exposing the germinating seed to low temperature. *Nature*, 138: 249.

156 Maksimov, N. A. 1929: *Selskokhozaistvennaia Gazeta*, 19 November.

157 Chouard, P. 1960: Vernalization and its relation to dormancy. *Annual Review of Plant Physiology*, 11: 191.

158 Lysenko, T. D. and Baskova, N. 1939: *Yarovizatsiia*, p. 4.

159 Joravsky, D. 1970: *The Lysenko Affair*, Cambridge, Mass.: Harvard University Press.

160 Lysenko, T. D. 1937: *Spornoye Voprosy Genetiki i Selektsii*. Lenin All Union Academy of Agricultural Sciences, p. 61.

161 Lysenko, T. D. 1947: *Spornoye Voprosy Genetiki i Selektsii*. Lenin All Union Academy of Agricultural Sciences, p. 47.

162 Kostriukova, K. Y. 1962: *Nauchnye Doklady Vyzshey Shkely* Seria Filosofii No. 1.

163 Muller, H. J. 1937: *Spornoye Voprosy Genetiki i Selektsii*. Lenin All Union Academy of Agricultural Sciences, pp. 143–4.

164 Gould, S. J. 1981: The most chilling statement. *Natural History*. 90: 14.

165 Dolgushin, B. 1949: Michurinske Prinstipy Selektsii i Semenovodstva Kulturnykh Rasteniy, pp. 22–3.

166 Lysenko, T. D. 1950: *Problemy Botaniki*, 1, 167.

167 Lysenko, T. D. 1957: *Agrobiologiya*, No. 6, 9.

168 Medvedev Zh. 1969: *The Rise and Fall of Lysenko*. New York: Columbia University Press.

169 Lysenko, T. D. 1949: *Agrobiologiya*, (Sel'khozgiz) 602.

170 Voronov, B. 1964: *Selskaya Zhizn'*, 25 November.

171 Gorodinsky, K. 1964: Facts versus fabrication. *Komsomolskaya Pravda*, 29 November.

172 Koldanov, V. I. 1954: *Lesnoye Khoizaistvo*, 3, 10.

173 Karapetian, V. K. 1948: *Agrobiologiya*, 4: 5.

174 Kuhn, T. S. 1970: *The Structure of Scientific Revolutions*, 2nd ed. Chicago: Chicago University Press.

175 Cohen, I. B. 1985: *Revolution in Science*. Cambridge, Mass.: Harvard University Press.

176 Summerlin, W. T., Miller, G. E. and Good, R. A. 1973: Successful tissue and organ transplantation without immunosuppression. *Journal of Clinical Investigation*, 52: 83a.

177 Medawar, P. B. 1976: The strange case of spotted mice. *New York Times Review of Books*, 15 April, p. 8.

178 McBride, G. 1974: The Sloan Kettering affair. Could it have happened anywhere? *Journal of the American Medical Association*, 229: 1391.

179 Hixson, J. 1976: *The Patchwork Mouse*. Garden City: Anchor Press/ Doubleday.

180 Goodfield, J. 1981: *Reflections on Science and the Media*. Washington, DC: American Association for the Advancement of Science, 81–5.

181 Culliton, B. J. 1983: Emory reports on Darsee's fraud. *Science*, 220: 936.

182 Kloner, R. A., De Boer, L. W. V., Darsee, J. R., Ingwal, J. S. and Braunwald, E. 1981: Prolonged abnormalities of myocardium salvaged by reperfusion. *American Journal of Physiology*, 240: 541.

183 Kloner, R. A., De Boer, L. W. V., Darsee, J. R., Ingwal, J. S. and Braunwald, E. 1981: Recovery from prolonged abnormalities of canine myocardium salvaged from ischemic necrosis by coronary reperfusion. *Proceedings of the National Academy of Science, USA*, 78: 7152.

184 Darsee, J. R. and Kloner, R. A. 1981: Dependency of location of salvageable myocardium on type of intervention. *American Journal of Cardiology*, 58: 702.

185 Culliton, B. J. 1983b: Coping with fraud. The Darsee case. *Science*, 220:31.

186 Darsee, J. R., Kirsch, C. M., Wynne, J. and Holman, B. L. 1981: Comparative analysis of infarct size by emission computed tomography with standard clinical criteria for myocardial infarction in humans. American Heart Association Monograph No. 82, IV: 222.

187 Darsee, J. R. and Heymsfield, S. B. 1981: Decreased myocardial taurine levels and hypertaurinuria in a kindred with mitral valve prolapse and congestive cardiomyopathy. *New England Journal of Medicine*, 304: 130.

188 Broad, W. J. 1982: Harvard delays in reporting fraud. *Science*, 215: 478.

189 Wallis, C. 1983: Fraud in Harvard lab. *Time Magazine*, 28 February.

190 McDona, K. 1983: Investigations find Harvard researcher manipulated data. *Chronicle of Higher Education*, 3 February, p. 6.

191 Soman, V. and Felig, P. 1980: Insulin binding to monocytes and insulin sensitivity in anorexia nervosa. *American Journal of Medicine*, 68: 66.

192 Wachslicht-Rodbard, H., Gross, H. A., Rodbard, D., Ebert, M. H. and Roth, J. 1979: Increased insulin binding to erythrocytes in anorexia nervosa. *New England Journal of Medicine*, 300: 882.

193 Broad, W. J. 1980: Imbroglio at Yale, I. Emergence of fraud. *Science*, 210: 38.

194 Broad, W. J. 1980: Imbroglio at Yale. II. A top job lost. *Science*, 210: 171.

195 Felig, 1981: Hearings before the Sub-committee on Investigations and Oversight of the Committee on Science and Technology US House of Representatives, Washington, US Government Printing Office.

196 Roark, A. C. 1980: Scientists question profession's standards amid accusations of fraudulent research. *Chronicle of Higher Education*, 2 September, F-6634.

197 Long, J. C., Aisenberg, A. C., Zamecnik, M. V. and Zamecnik, P. A. 1973: A tumor antigen in tissue culture derived from patients with Hodgkin's disease. *Proceedings of the National Academy of Science, USA*, 70: 1540.

198 Long, J. C. Zamecnik, P. C., Aisenberg, A. C. and Atkins, L. 1977: Tissue culture studies in Hodgkin's disease. Morphologic, cytogenetic, cell surface and enzyme properties. *Journal of Experimental Medicine*, 145: 1484.

199 Long, J. C., Dvorak, A. M., Quay, S. C., Stamatos, C. and Chi, S. Y. 1979: Reaction of immune complexes with Hodgkin's disease tissue culture. Radioimmunoassay and immunoferritin electron microscopy. *Journal of National Cancer Institute*, 62: 782.

200 Zamecnik, P. C. and Long, J. C. 1977: Growth of cultured cells from patients with Hodgkin's disease and transplantation into nude mice. *Proceedings of the National Academy of Science, USA*, 74: 754.

201 Long, J. C., Hall, C. L., Brown, C. A., Stamatos, C., Weitzman, S. A. and Carey, K. 1977: Binding of soluble immune complexes in serum of patients with Hodgkin's disease in tissue culture derived from the tumor. *New England Journal of Medicine*, 29: 295.

202 Harris, N. L., Gang, D. L., Quay, S. C., Poppema, S., Zamecnik, P. C., Nelson-Rees, W. A. and O'Brien, J. J. 1981: Contamination of Hodgkin's disease cell cultures. *Nature*, 289: 228.

203 Dickson, D. 1981: Contaminated cell lines. *Nature*, 289: 227.

204 Spector, M., O'Neel, S. and Racker, R. 1981: Regulation of phosphorylation of the subunit of the Ehrlich ascites tumor Na^+K^+ ATPase by a protein kinase cascade. *Journal of Biological Chemistry*, 256: 4219.

205 Spector, M., Pepinsky, R. B., Vogt, V. M. and Racker, E. 1981: A mouse homolog of the avian sarcoma virus src protein in a member of a protein kinase cascade. *Cell*, 25: 9.

206 Stein, M. D. 1981: Cornell retracts reports of kinase cascade. *Nature*, 293: 93.

207 Vogt, V. M., Pepinsky, R. B. and Racker, E. 1981: src protein and the kinase cascade. *Cell*, 25: 827.

208   Spector, M. and Winget, C. 1980: Purification of a manganese containing protein involved in photosynthetic oxygen evolution and its use in reconstituting an active membrane. *Proceedings of National Academy of Science, USA*, 77: 957.

209   Newmark, P. 1981: Hidden spectre. *Nature*, 293: 32.

210   Newmark, P. 1981: Biochemical cascades and carcinogenesis. *Nature*, 293: 93.

211   Webster, G. 1963: Sequence of reactions in the phosphorylation coupled to the oxidation of reduced cytochrome C. *Biophysical Biochemical Research Communications*, 13: 399.

212   Webster, G. and Green, D. E. 1964: The enzymes and the enzyme catalysed reactions of mitochondrial oxidative phosphorylation. *Proceedings of the National Academy of Science, USA*, 52: 1170

213   Racker, E. 1981: Warburg effect revisited: Merger of biochemistry and molecular biology. *Science*, 213: 303.

214   Yanchinski, S. 1981: The cancer cascade: Why the science is suspect. *New Scientist*, 92, 1 October: 22.

215   Wade, N. 1981: The rise and fall of a scientific superstar. *New Scientist*, 91, 24 September: 781.

216   Enomoto, K. and Lucas, Z. J. 1972. Mechanism of immunologic enhancement of kidney grafts in rats. *Surgery Forum*, 24: 269.

217   Morris, R. and Lucas, Z. J. 1971: Immunologic enhancement of rat kidney grafts. Evidence for peripheral action of homologous antiserum. *Transplantation Proceedings*, 3: 697.

218   Enomoto, K. and Lucas, Z. J. 1973. Role of the spleen in immunologic enhancement of kidney grafts in rats. *Advances in Experimental Medicine and Biology*, 29: 419.

219   Fortney, M. T. 1981: Stanford research case settled out of court. *Peninsula Time Tribune*, 19 January.

220   Lucas, Z. J. and Walker, S. M. 1974: Cytotoxic activity of lymphocytes. III. Standardization of measurements of cell mediated lysis. *Journal of Immunology*, 113: 209.

221   Currens, R. C. 1977: Trip report–Review of research reporting. Published by Stanford University, 29 August.

222   Max, B. 1984: Ethics, ambiguity and authorship. *Trends in Pharmacological Sciences*, May, p.180.

223   Paradies, H. 1968: A method for crystallization of serine-transfer RNA. Cocrystallization of t-RNA with cadmium and copper ion in water-dioxane. *FEBS Letters*, 2: 112.

224   Paradies, H. and Sjoquist, H. 1970: Crystallographic study of valine t-RNA from yeast. *Nature*, 303: 196.

225   Zimmer, B., Paradies, H., and Werz, G. 1977: Electron microscopic studies on microcrystals of d-ribulose- 1,5-biphosphate carboxylase from *Dasycladus clavaeformi*. *Biophysical Biochemical Research Communications*, 74: 1496.

226 Paradies, H. 1979: Crystallization of coupling factor 1 (CF_1) from spinach protoplasts. *Biophysical Biochemical Research Communications*, 91: 685.

227 Hendrickson, W. A., Strandberg, B. E., Liljas, A., Amzal, L. and Lattman, E. E. 1983: True identity of a diffraction pattern attributed to valyl t-RNA. *Nature*, 303: 195.

228 Newmark, P. 1983: Disputed X-ray data unresolved. *Nature*, 303: 197.

229 Paradies, H. 1971: Polymorphism of serine specific transfer ribonucleic acid. *European Journal of Biochemistry*, 18: 530.

230 Paradies, H. 1974: Isolation and crystallization of valyl- tRNA synthetase from *E. coli* MRE 600. *Journal of Biochemistry (Tokyo)*, 76: 655.

231 Paradies, H. 1983: A reply from Paradies. *Nature*, 303: 196.

232 Clarck, B. F. C., Doctor, B. P., Holmes, K. C., Klug, A., Marcker, K. A., Morris, S. J. and Paradies, H. H. 1968: Crystallization of transfer RNA. *Nature*, 219: 1222.

233 Purves, J. 1981: Cerebral blood flow and metabolism in sheep fetus. In *Advances in Physiological Sciences*, Oxford: Pergamon Press.

234 Purves, J. 1981: In Bristol now. *Nature*, 294: 511.

235 Mead, M. 1933: *Coming of Age in Samoa. A Psychological Study of Primitive Youth for Western Civilization*. New York: Blue Ribbon Books.

236 Freeman, D. 1983: *Margaret Mead and Samoa: The Making and Unmaking of an Anthropological Myth*, Cambridge, Mass.: Harvard University Press.

237 Rensberger, B. 1983: Margaret Mead. The nature nuture debate. *Science 83* (April): 28–37.

238 Goodenough, W. 1983: Margaret Mead and cultural anthropology. *Science*, 220: 906.

239 Hanlon, J. 1974: Top food scientist published false data. *New Scientist*, 64 (7 November): 436.

240 Hutchinson, J. 1975: A sense of proportion. *New Scientist*, 65 9 January: 96.

241 Borlaug, N. E. and Anderson, G. R. 1975: Defence of Swaminathan. *New Scientist*, 65 30 January: 280.

242 Kar, S. 1975: Ethics in scientific journalism. *Science and Culture*, 41: 43.

243 Briggs, R. and King, T. J. 1962: Transplantation of living nuclei from blastula cells into enucleated frog cells. *Proceedings of the National Academy of Science*, 38: 455.

244 Hoppe, P. C. and Illmensee, K. 1977: Microsurgically produced homozygous–diploid uniparental mice. *Proceedings of the National Academy of Science, USA*, 74: 5657.

245 Illmensee, K. and Hoppe, P. C. 1981: Nuclear transplantation in *Mus musculus*: Developmental potential of nuclei from preimplantation embryos. *Cell*, 23: 9.

246 Marx, J. L. 1983: Bar Harbor investigation reveals no fraud. *Science*, 220: 1023.

247 Marlatt, G. A. 1983: The controlled drinking controversy. *American Psychologist*, 38: p.1097.

248 Jellinek, E. M. 1960: *The Disease Concept of Alcoholism*. New Brunswick, NJ: Hillhouse Press.

249 Davies, D. L. 1962: Normal drinking in recovered alcoholic addicts. *Quarterly Journal of Studies on Alcohol*, 23: 94.

250 Sobell, M. B. and Sobell, L. C. 1973: Individualized behavior therapy for alcoholics. *Behavior Therapy*, 4: 49.

251 Sobell, M. B., and Sobell, L. C. 1978: *Behavioral Treatment of Alcohol Problems*. New York: Plenum Press.

252 Miller, W. R. and Hester, R. K. 1980: Treating the problem drinker. Modern approaches. In W. R. Miller (ed.) *The Addictive Behaviors*, Oxford: Pergamon Press.

253 Sobell, M. B., and Sobell, L. C. 1982: Controlled drinking. A concept coming of age. In K. R. Blanstein and J. Polivy (eds) *Self Control and Self Modification of Emotional Behavior*, New York: Plenum Press.

254 Pendery, M. L., Maltzman, I. M. and West, L. J. 1982: Controlled drinking by alcoholics? New findings and reevaluation of a major affirmative study. *Science*, 217: 169.

255 Sobell, M. B. and Sobell, L. C. 1984: The aftermath of heresy, a response to Pendery *et al's* (1982) critique of individualized behavior therapy for alcoholics. *Behavior Research and Therapy*, 22: 413.

256 Sobell, M. B., and Sobell, L. C. 1984: Under the microscope yet again: A commentary on Walker and Roach's critique of the Dickens Committee enquiry into our research. *British Journal of Addiction*, 79: 157.

257 Davies, D. L. 1981: Foreword. In N. Heather and I. Robertson (eds) *Controlled Drinking*, London: Methuen.

258 Nelson–Rees, W. A., Flandermeyer, R. R. and Daniels, D. W. 1980: T-1 cells are HeLa cells and not of normal human kidney origin. *Science*, 209: 719.

259 Furcinitti, P. S. and Todd, P. 1979: Gamma rays. Further evidence for lack of threshold dose for lethality to human cells. *Science*, 206: 475.

260 Broad, W. J. 1980: The case of unmentioned malignancy. *Science*, 210: 1229.

261 Todd, P., Geraci, J. P., Furcinitti, P. S., Rossi, R. M., Mikage, F., Theus, R. B. and Schroy, C. B. 1978: Comparison of the effects of various cyclotron produced fast neutrons on the reproductive capacity of cultured human kidney (T-1) cells. *International Journal of Radiation Oncology Biology and Physics*, 4: 1015.

262 Wheeler, W. E, Noller, C. H. and White, J. L. 1981: Effect of rate of reactivity on calcitic limestone and level of calcium addition on

utilization of high concentrate diets by beef steer. *Journal of Animal Science*, 53 (Suppl. 1): 444.

263 Wheeler, W. E., Noller, C. H., White, J. L. and Hruska, R. L. 1980: Influence of limestone on the performance of beef steers fed corn silage corn grain based diets. *Journal of Animal Science*, 51 (Suppl. 1): 409.

264 Eng, K. 1982. Limestone research: where do we go from here? *Feedstuffs*, 54: 23.

265 Barber, B. 1961: Resistance by scientists to scientific discovery. *Science*, 134; 596.

266 Weiner, J. S. 1955: *The Piltdown Forgery*. London: Oxford University Press.

267 Keith, A. 1925: *The Antiquity of Man*. 2nd edn London: Williams and Norgate, 667.

268 Oakley, K. P. and Hoskins, C. R. 1950: New evidence on the antiquity of Piltdown man. *Nature*, 165: 379.

269 Weiner, J. S., Oakley, K. P. and Le Gros Clark, W. E. 1953: The solution of the Piltdown problem. *Bulletin of the British Museum, (Natural History) Geology*, 2: 139.

270 Woodward, A. S. 1916: *Geological Magazine* 3: 477 (quoted by Weiner (1955).

271 Gould, S. J. 1980: The Piltdown conspiracy. *Natural History*, 8: 8.

272 Gould, S. J. 1980: *The Panda's Thumb*. New York and London: W. W. Norton & Co.

273 Harrison Mathews, L. 1981: Piltdown man. The missing link. *New Scientist*, 90, 30 April: 280.

274 Watkins, R. S., Hoyle, F., Wickramasinghe, N. C., Watkins, J., Rabilizirow, R. and Spetner, L. M. 1985: Archeopteryx – a photographic study. *British Journal of Photography*, 132: 264.

275 Watkins, R. S., Hoyle, F., Wickramasinghe, N. C., Watkins, J., Rabilizirow, R. and Spetner, L. M. 1985: Archeopteryx – a further comment. *British Journal of Photography*, 132: 358.

276 Watkins, R. S., Hoyle, F., Wickramasinghe, N. C., Watkins, J., Rabilizirow, R. and Spetner, L. M. 1985: Archeopteryx – further evidence. *British Journal* of *Photography*, 132: 468.

277 Williams, N. 1985: Fraudulent feathers. *Nature*, 314: 210.

278 Vines, G. 1985: Strange case of Archeopteryx 'fraud'. *New Scientist*, 105, March 14:9

279 Owen, R. 1863: *Philosophical Transactions of the Royal Society*, 153: 33–47.

280 Wendt, H. 1970; *Before the Deluge*. London: Paladin/Granada.

281 Clairemont-Ganneau, Ch. 1885: *Les Frauds Archéologiques en Palestine*. Paris; E. Leroux.

282 Harry, M. 1914: *La Petite Fille de Jerusalem* (Hebrew translation, 1975). Tel Aviv: Levinson Press.

283 Stevens, M. and Foote, D. 1984: A master forger exposed. *Newsweek*, 29 January, p. 49.

284 Hagstrom, W. 1974: Competition in science. *American Social Review*, 39: 1.

285 Wheelock, E. F. 1980: Plagiarism and freedom of information laws. *Lancet*, 1: 826.

286 Alsabti, E. A. 1979: Tumor dormancy: (a review). *Journal of Cancer Research in Clinical Oncology*, 95: 209.

287 Alsabti, E. A. 1979: Tumor dormancy: (a review). *Neoplasna*, 26: 351.

288 Broad, W. J. 1980: Would-be academician pirates paper. *Science*, 208: 1438.

289 Broad, W. J. 1980: An outbreak of piracy in the literature. *Nature*, 285: 429.

290 Alsabti, E. A., Ghalib, O. N. and Salem, M. H. 1979: Effect of platinum compounds on murine lymphocyte mitogenesis. *Japanese Journal of Medical Science and Biology*, 32: 53.

291 Wierda, D. and Pazdernik, T. L. 1979: Suppression of spleen lymphocyte mitogenesis in mice injected with platinum compounds. *European Journal of Cancer*, 15: 1013.

292 Shishido, A. 1980: One journal disowns plagiarism. *Nature*, 286: 437.

293 Shishido, A. Retraction notice. *Japanese Journal of Medical Science and Biology*, 33: 4.

294 Alsabti, E. A. 1979: Serum lipids in hepatoma. *Oncology*, 36: 11.

295 Yoshida, T., Okazaki, N., Yoshuno, M. and Araki, E. 1977: Diagnostic evaluation of serum lipids in patients with hepatocellular carcinoma. *Japanese Journal of Clinical Oncology*, 7: 15.

296 Alsabti, E. A. 1979: Lymphocyte transformation in patients with breast cancer and the effect of surgery. *Japanese Journal of Experimental Medicine*, 49: 101.

297 Watkins, C. H. 1973: The effects of surgery on lymphocyte transformation in patients with breast cancer. *Clinical and Experimental Immunology*, 14: 69.

298 Abelson, P. H. 1982: Excessive zeal to publish. *Science*, 218: 953.

299 Ferguson, M. W. J. 1981: Extrinsic microbial degradation on the alligator eggshell. *Science*, 214: 1135.

300 Ferguson, M. W. J. 1981: Increased porosity of the incubating alligator eggshell caused by extrinsic microbial degradation. *Experientia*, 37: 252.

301 Lock, S. 1984: Repetitive publication: a waste that must stop. *British Medical Journal*, 288: 661.

302 International Committee of Medical Journal Editors (12 names) *British Medical Journal*, 1985: 722.

303 Ginsburg, I. 1982: Can the editorial boards of ASM journals afford to be infallible? *ASM News*, 48: 555.

304 Davis, B. D., Stottmeier, K. D., Teichman, N., Novotny, A., Chen, J.-S. and Kabat, E. A. 1983: Should ASM journals publish letters to the editors? A range of opinions. *ASM News*, 49: 185.

305 Shands, J. W. 1983: Editor in chief of IAI responds. *ASM News*, 49: 186. -

306 Gray, P. 1983: Fakes that have skewed history. *Newsweek*, 16 May, 25.

307 Magnuson, E. Hitler's forged diaries. *Time Magazine*, 16 March, 14.

308 Rapp, F. 1981: Attack on science: erosion from within. *ASM News*, 47: 193.

309 Budiansky, S. 1985: New ways of shading truth. *Nature*, 315: 447.

310 Greenberg, D. 1982: Fraudian analysis. *New Scientist*, 2 September, 95: 643.

311 van Valen, L. 1976: Dishonesty and grants. *Nature*, 261: 2.

312 Lodish, H. 1982: Validity of scientific data – The responsibility of the principal investigator. *Trends in Biochemical Sciences*, March, 82.

313 Potkin, S. G., Cannon, H. E., Murphy, D. L. and Wyatt, R. J. 1978: Are paranoid schizophrenics different from other schizophrenics? *New England Journal of Medicine*, 298: 61.

314 Pajari, K. J. 1978: Momoamine oxidase in schizophrenia. *New England Journal of Medicine*, 298: 1150.

315 Berger, P. A., Ginsberg, R. A., Barchas, J. D., Murphy, D. L. and Wyatt, R. J. 1978: Platelet monoamine oxidase in chronic schizophrenic patients. *American Journal of Psychiatry*, 135: 95.

316 Potkin, S. G., Cannon, H. E., Murphy, D. L. and Wyatt, R. J. 1978: Monsamine oxidase in schizophrenia (correspondence) *New England Journal of Medicine*, 298: 1151.

317 Relman, A. 1978: Schizophrenia and publication. *New England Journal of Medicine*, 298: 1152.

318 Wyatt, R. J. and Murphy, D. L. 1978: Schizophrenia and publication: epilogue. *New England Journal of Medicine*, 299: 424.

319 Broad, W. J. 1981: The publishing game: getting more for less. *Science*, 211: 1137.

320 Culliton, B. J. 1978: Scientists dispute book's claim that human clone has been born. *Science*, 199: 1314.

321 Broad, W. J. 1981: Saga of boy clone ruled a hoax. *Science*, 211: 902.

322 Bromhall, D. 1978: The great cloning hoax, *New Statesman*, 2 June, 734.

323 Broad, W. J. 1982: Publisher settles suit, says clone book is a fake. *Science*, 216: 391.

324 Sun, M. 1983: Scientists settle cell line dispute. *Science*, 220: 394.

325 Wade, N. 1980: University and drug firm battle over billion-dollar genes. *Science*, 209: 1492.

326 Nelkin, D. 1982: Intellectual property: the control of scientific information. *Science*, 216: 704.

327 Lappe, M. 1975: Accountability of science. *Science*, 187: 696.

328 Dowie, M. Foster, D., Marshall, C., Weir, D. and King, J. 1982: The illusion of safety. *Mother Jones*, June: 38–49.

329 Budiansky, S. 1982: Drug testing lab was 'shambles'. *Nature*, 302: 730.

330 Marshall, E. 1983: Federal court finds IBT officials guilty of fraud. *Science*, 222: 488.

331 Marshall, E. 1983: The murky world of toxicity testing. *Science*, 220: 1130.

332 Marshall, E. 1983: Pesticide office demands new safety studies. *Science*, 221: 442.

333 Holden, C. 1979: FDA tells senators of doctors who fake data in clinical drug trial. *Science*, 206: 432.

334 Budiansky, S. 1983: Food and drug data fudged. *Nature*, 302: 560.

335 Insight Team (of *The Sunday Times*) 1979: *Suffer the Children. The Story of Thalidomide.* New York: Viking Press.

336 Harmon, W. E., Cohen, H. J., Schneeberger, E. E. and Grupe, W. E. 1979: Chronic renal failure in children treated with methyl-CCNU. *New England Journal of Medicine*, 300: 1200.

337 Kubie, L. S. 1953: Some unresolved problems of the scientific career. *American Scientist*, October: 596; January 1954: 104.

338 Kubie, L. S. 1958: *Neurotic Distortion of the Creative Process*, Terrance, Ca.: The Noontide Press.

339 Davis, B. D. 1983: Neo-Lysenkoism, IQ and the press. *The Public Interest*, Fall 41.

340 Zinder, N. and Meyer, A. 1983: Fraud in science, a scientist's view. *Science*, 83: 94.

341 Ziman, J. M. 1970: Some pathologies in the scientific life. *Nature*, 227: 996.

342 Price, A. R. 1983: Dealing with scientists who cheat. *Chemical and Engineering News*, June 13, 68.

343 Wade, N. 1983: What science can learn from science fraud. *New Scientist*, 99, 28 July: 273.

344 Chalk, R. and van Hippel, F. 1979: Due process for dissenting whistle-blowers. *Technological Reviews*, 8: 48.

345 Holden, C. 1980: Scientist with unpopular data loses job. *Science*, 210: 749.

346 Westin, A. F. 1980: *Whistle Blowing – Loyalty and Dissent in Corporations.* New York: McGraw–Hill.

347 Edsall, J. T. 1981: Two aspects of scientific responsibility. *Science*, 212: 11.

348 Executive Council Association of American Medical Colleges 1982: Maintenance of high ethical standards in the conduct of research. *Journal of Medical Education*, 57: 896.

349 Culliton, B. J. 1983d: AAAC speaks on coping with fraud. *Science*, 217:226.

350 Broad, W. J. 1982: Yale announces plan to handle charges of fraud. *Science*, 218: 37.

351 Broad, W. J. 1982: NIH grapples with misconduct. *Science*, 217: 227.

352 Dixon, B. and Holister, G. 1984:*Ideas of Science*, Oxford: Basil Blackwell, 53.

Index